P9-CEG-342

DATE DUE

DISCARD

What the Nose Knows

What the Nose Knows

The Science of Scent in Everyday Life

Avery Gilbert

CROWN PUBLISHERS
NEW YORK

Copyright © 2008 by Avery N. Gilbert

Published in the United States by Crown Publishers, an imprint of the
Crown Publishing Group, a division of Random House, Inc., New York.
www.crownpublishing.com

CROWN and the Crown colophon are registered trademarks
of Random House, Inc.

Grateful acknowledgment is made to the following for permission to reprint
previously published material:

Global Cosmetic Industry Magazine: Excerpts from "Scent and the City: Q & A
with Dennis Keogh, vice president of marketing, Lancaster Group" (April 2004).
Reprinted by permission of Global Cosmetic Industry Magazine.

National Review, Inc.: Excerpt from "Smellie on Seventh Avenue,"
by Joan Didion (January 30, 1960), copyright © 1960 by National Review, Inc.
Reprinted by permission of National Review, Inc., 215 Lexington Avenue,
New York, NY 10016.

Library of Congress Cataloging-in-Publication Data

Gilbert, Avery N.
    What the nose knows : the science of scent in everyday life /
Avery Gilbert.—1st ed.
    Includes bibliographical references and index.
    1. Smell—Popular works.   2. Odors—Popular works.   I. Title.
    QP458.G535 2008
    612.8'6—dc22   2007050009

ISBN 978-1-4000-8234-6

Printed in the United States of America

Design by Lauren Dong

10   9   8   7   6   5   4   3   2   1

First Edition

To Susie

If you are ambitious to found a new science, measure a smell.

—ALEXANDER GRAHAM BELL, 1914

"They've injected us! We're off on our fantastic
    journey through the human body!"

"Just think! In less than one minute, we have a
    rendezvous with our olfactory destiny!"

"Isn't that a rather pretentious statement, Doctor?"

"All right! YOU think of a more romantic way
    to say we're going up a nose!"

—"FANTASTECCH VOYAGE," *Mad Magazine*, 1967

Contents

Introduction   *xi*

CHAPTER 1   Odors in the Mind   *1*
Making sense of scents

CHAPTER 2   The Molecules That Matter   *25*
The key ingredients of nature's chemistry set

CHAPTER 3   Freaks, Geeks, and Prodigies   *48*
The wide range of olfactory ability

CHAPTER 4   The Art of the Sniff   *74*
How the brain edits what the nose detects

CHAPTER 5   A Nose for the Mouth   *91*
Cooking, eating, and evolution

CHAPTER 6   The Malevolence of Malodor   *111*
When bad smells happen to good people

CHAPTER 7   The Olfactory Imagination   *126*
Smell and creative genius

CHAPTER 8   Hollywood Psychophysics   *147*
Scent and cinema

CHAPTER 9   Zombies at the Mall   *170*
Leading consumers by the nose

CHAPTER 10   Recovered Memories   *189*
Proust's soggy madeleine

CHAPTER 11   The Smell Museum   *205*
Preserving odors from the past

CHAPTER 12   Our Olfactory Destiny   *225*
The future of smelling

Acknowledgments   *239*
Notes   *241*
Index   *281*

Introduction

BEING A SMELL EXPERT has its ups and downs. In my professional life I've attended secret perfume planning meetings in corporate suites with spectacular views of Central Park. But I've also sat at a conference table and sniffed defrosted samples of used feminine hygiene products.

I've traveled to London, Zürich, Paris, and Cannes, staying at the best hotels and eating at the finest restaurants. I've also traveled to Cape Girardeau, Missouri, to evaluate the scent of carefully aged cat feces.

I have air-kissed fashion celebrities and sniffed the scalps of elderly ladies being shampooed in a mock salon.

I was among the first people to smell Elizabeth Taylor's *White Diamonds,* but also one of the first to sniff purified 3-methyl-2-hexenoic acid—the aromatic essence of ripe, unwashed armpits.

These experiences are not unusual in the fragrance industry: people there sniff for a living and create scents for everything from perfume to kitty litter. What is unusual is that I'm a sensory psychologist, trained in evolutionary theory, animal behavior, and neuroscience. I'm a rational, evidence-based guy working in the most frothy, fashion-driven, marketing-heavy business outside of Hollywood.

The sense of smell portrayed in the mainstream media ("Seven Ways to Drive Him Wild with Your Perfume!") is far different from the way scientists see it ("Multivariate Analysis of Odorant-Induced Neural Activity in the Anterior Piriform Cortex"). The magazine version—breezy and chatty—sails merrily past new discoveries just emerging from the laboratory. The official scientific version—formal, dense, and dry—hides some very cool new stories.

I know people are fascinated by the hows and whys of odor perception. When they find out I'm an expert, they bombard me with questions. The answers are often weirder than they could have imagined. The new science of smell is making us rethink everything from wine tasting to Smell-O-Vision. So it's time for a fresh look at odor perception and how it plays out in popular culture.

Where to begin? With a simple question: How many smells are there? The answer leads to psychology ("How do you count smells?"), technology ("How do you take apart a complex odor?"), and secrets of the trade ("How do you become a perfumer?").

In the chapters that follow, I take up other simple questions—What makes for a good sense of smell? Do bad odors make us sick? Can subliminal smells make us do things against our will?—and follow them to some strange and unexpected places. Welcome to my world. Breathe deep and enjoy.

What the Nose Knows

CHAPTER I

Odors in the Mind

It is very obvious that we have very many different kinds of smells, all the way from the odor of violets and roses up to asafetida. But until you can measure their likenesses and differences you can have no science of odor.

—ALEXANDER GRAHAM BELL, 1914

No satisfactory classification of odours can be given.

—*Encyclopaedia Britannica*, 1911

HOW MANY SMELLS ARE THERE? IT'S AN ODD QUESTION, but give it some thought. Mentally flip through the pages of your personal smell catalog. You find burnt toast, shaving cream, Grandma's kitchen, and pine trees. There's the weird glue in the binding of that pocket-size Latin/English dictionary from high school. With a little effort you can come up with a lot of smells, but putting a number to them is difficult. How does one count the odors of a lifetime, much less all the odors in the world?

Some people aren't daunted by the task: they simply estimate. Better yet, they pass along estimates made by others. Journalists like to say that we can smell thirty thousand different odors. The New Age guru Michael Murphy cites this figure in *The Future of the Body* (1992): "According to the calculations of one [fragrance] manufacturer, an expert can distinguish more than 30,000 nuances of scent." Murphy, in turn, got the number from Vitus Dröscher, a German pop science writer (1969): "A perfume manufacturer has worked out that a real expert must distinguish at least thirty thousand nuances of scent." Dröscher doesn't provide a source. Perhaps it was in *Science Digest* (1966): "Industrial chemists have identified some 30,000 different smells." Unfortunately, the magazine didn't provide a source either. What this proves, I suppose, is that dubious facts thrived in the media long before the Internet.

One would like to think that smell scientists have a better grasp of the matter, and indeed they prefer to quote a different estimate. When Linda Buck and Richard Axel won the 2004 Nobel Prize for discovering the olfactory receptors, the Nobel Foundation issued a press release. It noted that people recognize and remember "about 10,000 different odours," a figure the Swedish publicists took from the prize winners themselves. Surely that's a number we can take to the bank. But the number 10,000 didn't originate with Buck and Axel: it had been tossed about for years by other scientists. Something about it had always bothered me—why such a nice fat round number? Why was there no date of discovery? And, strangest of all, why did nobody take credit for it?

If you try to track down the mysterious number 10,000 to its original source in the scientific literature, you are in for an adventure; like walking a maze, dead ends abound. For example, I began with a paper in *Behavioral Ecology* (2001), which I followed to another in *Trends in Genetics* (1999), which in turn didn't provide a source. I started again, this time with a prominent smell researcher, the Brown University psychologist Trygg Engen. In 1982 he wrote, "Some have claimed that an untrained person can identify by label at least 2,000 odors and an expert can identify as many as 10,000." Engen credits this claim to R. H. Wright, Canada's most famous smell scientist. Wright seemed a likely source, at least until I read what he actually wrote, back in 1964: "[I]t seems likely that the average person would have no trouble in distinguishing between several thousand odours, and an experienced authority in the field has claimed the ability to recognize well over ten thousand. Still another has simply said the number is apparently unlimited." Wright goes on to say, "It would be an interesting exercise to design an experiment to verify these estimates." Ooof! So Wright didn't discover any number at all—he just passed along what he had heard and Professor Engen repeated it. These eminent smell experts remind me of kids at summer camp passing along ghost stories.

I was beginning to think I'd never find the source of the magic number 10,000, when I found it once more in a 1999 food chemistry textbook. From there I followed it to a 1966 paper, and then to a paper published in 1954 by researchers from the Arthur D. Little, Inc., consulting company. At a scientific conference the previous year, the Little folks presented a paper titled "An information theory of olfaction." Their goal was to place numerical limits on odor perception. They said "There are experts who affirm that it is possible to recognize at least 10,000 odors," a figure they used in their mathematical subsequent analysis. The name of their expert was buried in a footnote: he was Ernest C. Crocker, a chemical engineer and 1914 MIT graduate who, not coincidently, was also an employee of Arthur D. Little, Inc.

Back in 1927, Crocker and another Little chemist named Lloyd F. Henderson were struggling for an objective way to classify odors. They settled on a method in which an odor was rated by how strongly (on a scale of 0 to 8) it resembled each of four elementary odor sensations. Given the mathematics of their rating system, it was theoretically possible to discriminate 9^4 or 6,561 different odors. The math is watertight, but the outcome is highly dependent on the initial assumptions. Had Crocker and Henderson used, say, five elementary sensations and a 0-to-10 rating scale, the estimate would have been 11^5 or 161,051 different odors. (Harvard psychologist Edwin Boring was a fan of the new system, but he believed the rating scale should have fewer steps. He did some calculations and decided that the number of distinguishable smells was somewhere between 2,016 and 4,410.) Discussing this work years later, Ernest Crocker generously rounded up the estimate to 10,000 odors. His colleagues took the number and ran with it.

In the end, it appears that no one has ever attempted to count how many smells there are in the world. Estimates of odor diversity lead either to a dead end or to Ernest C. Crocker. The comfortable, often-cited figure of 10,000 smells is, from a scientific perspective, utterly worthless.

WHY DOES IT MATTER exactly how many smells there are? Suppose we want to build a device that can reproduce every possible odor. (This is a popular fantasy. As a kid you may have scratch-and-sniffed your way through *Mickey Mouse and the Marvelous Smell Machine*.) A pair of industrial engineers once looked into how many distinct odors it would take to create a lifelike smellscape in virtual reality. They settled on a figure of 400,000. (This number has no more basis in fact than 10,000 or 30,000; its ultimate source is an obscure Japanese technical publication). Four hundred thousand is a staggeringly large number, but it sounds reasonable to engineers who use 16.7 million colors per pixel in visual displays for VR goggles. The trouble is that an engineer's solution doesn't always correspond to how the brain solves the problem.

The human eye detects tiny differences in color; across the visible spectrum we are capable of millions of such discriminations. Yet when it comes to naming categories of color, there is nearly universal agreement that only a half-dozen are needed to cover the range of human perception. People in all cultures get by very well with white, black, red, green, yellow, and blue. (Esoteric hues such as ecru and mauve occur mainly in clothing catalogs.) The physical spectrum of visible light is continuous; the stripes of the rainbow are created in our head. We give color names to these few categories.

This simplification of sensory input is a general feature of the brain, a phenomenon psychologists call categorical perception. In hearing, categorical perception helps us carve the continuous dimension of pitch into the individual notes of the musical scale, or the sonic blur of vowel sounds into a distinct *a* or *e*. Perhaps we shouldn't obsess about the number of odors. Our question should be, How many natural odor categories are there, and how do our nose and brain simplify the world?

The Art of the Achievable

I grew up in Davis, California, amid the smells of agriculture. Our house, when we first moved there in 1962, was near vast tomato fields; walking through them brought up the sharp, funky smell of the vines. The approach of a new school year was signaled by the heavy, stewed smell of tomatoes being cooked into ketchup at the Hunt-Wesson plant a mile upwind. My buddies and I played on mountains of newly baled alfalfa, stacked high and smelling grassy-sweet. The playground at the Valley Oak Elementary School offered the hot-metal smell of monkey bars and the dusty, sour resin of tan-bark underfoot. The water from the sprinklers in the town park had a musty tinge to it. The office of the *Davis Enterprise,* where I rolled copies for my afternoon paper route, was saturated with the smell of fresh ink, newsprint, and rubber bands. In grade school my class toured the Spreckles refinery, where truckloads of sugar beets were turned into pure white sugar, a magical transformation dimmed by the suffocating scent of dark molasses that hung over the place.

We moved to Davis because my father joined the faculty of the philosophy department at the university there. Davis was originally the agricultural field station for the main university campus at

Berkeley. Set in the hot, flat, and fertile Sacramento Valley, but near the cooler hills of Napa and Sonoma, UC Davis came into its own in the 1960s, when it added a law school and a medical school. At the same time, researchers in the Department of Viticulture and Enology got the ball rolling for what became the California wine-making revolution. They measured microclimates and soil composition, developed new grape varieties, and invented cold fermentation and other wine production techniques. Davis graduates who took courses in wine-tasting and wine-making are now among the world's leading vintners. As part of this effort, UCD researchers took up the sensory analysis of wine. Their challenge was to apply objective methods to one of the more rarefied arenas of subjective opinion: wine-tasting.

One of this group was Professor Ann Noble, a chemist and sensory specialist. Among her interests was identifying volatile chemical compounds in wine. These substances create the characteristic aroma of grape varieties such as Cabernet Sauvignon or Riesling, and also the off-odors found in poorly made wines. Noble hoped to link aroma chemistry to broader grape-growing and wine-making factors.

Noble's approach to aroma was practical and effective. Not for her the pretensions of the wine snob, epitomized by Paul Giamatti's character in the movie *Sideways,* who sticks his nose in a wineglass and says, "I'm getting strawberries, some citrus . . . passionfruit, just the faintest soupçon of asparagus, and, like, a nutty Edam cheese." To better understand California Cabernets, Noble and her colleague Hildegard Heymann selected wines from seven regions around the state (Napa, Sonoma, Alexander Valley, Santa Ynez, etc.). The wines were rated by enology students who came up with their own simple descriptive terms: berry, bell pepper, eucalyptus, and so on. Reference samples were created by doctoring a neutral "base" wine. To represent berry, for example, one-half teaspoon of raspberry jam and one half of a frozen blackberry were added to a half-cup of wine; after soaking for ten minutes, the blackberry was removed. For the soy/prune standard, the basic wine was spiked with a quarter-cup of canned prune juice and seventeen drops of Kikkoman soy sauce.

Armed only with their noses and a nine-point rating scale, the students sniffed and sipped their way to an enormous pile of data. (It filled a metaphorical filing cabinet thirteen descriptors wide, twenty-one Cabernets tall, and thirteen judges deep.) With a computer program, Noble and Heymann extracted a small number of sensory dimensions and placed each wine at a precise location on them. They could now visualize the smell and taste relationships between the samples. Their conclusions: "Younger vines and/or vines from cooler areas tend to produce more intensely vegetative wines. Conversely, wines from older vines and/or warmer areas tend to have higher ratings for berry aroma, fruit flavor by mouth, and vanilla aroma." By quantifying wine aromas with a system of practical description, they discovered the vineyard conditions that produce them. Alexander Graham Bell would have been pleased.

WHAT EARNED Ann Noble and her colleagues a place in the annals of smell classification was their Wine Aroma Wheel, published in 1984. The wheel was a visually pleasing presentation of a standardized wine aroma vocabulary. With twelve categories and ninety-four descriptive terms, it covers the aroma of any wine, regardless of grape or geographic origin. What makes the wheel just as useful to novices as to connoisseurs are the do-it-yourself reference standards. Simple kitchen chemistry lets anyone create and experience the standards for himself. Noble is critical of commercial wine-tasting kits; she believes that vials of flavor essence are chemically unstable and tend to degrade quickly. This spurred her to create reference standards that can be "prepared using foodstuffs available throughout the world during most seasons."

The Wine Aroma Wheel resembles a dartboard: three concentric circles divided pizzalike into a dozen slices of varying width. On the innermost circle, the pointed end of each wedge is an aroma category, such as fruity. In the middle circle, the wedge may be split into subcategories such as citrus, berry, or tree fruit. On the outer circle are specific materials, examples of each aroma subcategory. Thus you can follow the fruity wedge through the berry subslice to the outer circle; there you'll find blackberry, raspberry, strawberry, and blackcurrant. The beauty of Noble's wheel is that it links sensory concepts to actual everyday stuff—it connects Riesling to raspberries. Wheel in hand, it is possible to sniff your way to sensory enlightenment. This commonsense approach lets anyone grasp the esoteric inner-wedge category *microbiological* and its arcane subdivision *lactic*. It's only baffling until you sniff the examples: yogurt and sauerkraut. Then it clicks. The wheel even demystifies the wine-snob term *wet dog*: it's an example of *sulfur* aromas in the *chemical* category (along with skunk, cabbage, and burnt match).

There is no place on the wheel for the wine critic's gaseous adjectives. You will find "orange blossom" and "black olive" and the less flattering "soapy" and "cooked cabbage." But you will not find "an impertinent little Pinot Noir" or a "flabby, overripe Cabernet Franc," à la Miles Raymond. These free-form prose poems say more about a wine lover's pretensions than about the character of the wine. To use the wheel, all you need is a glass and grocery store.

A PRACTICAL SMELL classification for beer was created in the 1970s by a Danish flavor chemist named Morten Meilgaard. His Beer Flavor Wheel has now been adopted worldwide. It uses fourteen categories and forty-four sensory terms to describe the smell and taste of any style of beer—lager, ale, or stout. Most of the descriptors deal with aroma; others involve taste (bitter for hops; sweet for malt) and sensory factors like carbonation. Meilgaard's system includes reference standards, but unlike Noble's wine wheel (which was inspired by it), one needs access to pure chemicals to create them. For example, to mimic the "papery" aroma of oxidized beer, one doses a pitcher of beer with *trans*-2-nonenal.

A brewer's best friend is his nose. Desirable aromas tell him when the product is on target. By identifying off-smells in the product, a brewer can correct problems in the brewing process. For example, the smell of wet newspaper indicates that a beer has oxidized. Sunlight-damaged beer has a skunky smell. (Many years ago, Corona beer from Mexico was poorly made and oxidized easily. The acid in a slice of lime was an effective way to chemically neutralize the off-odor. Today Corona is made as well as any beer in the world, but the lime tradition lives on.)

Meilgaard's beer system is less satisfying for the lay drinker than the Wine Aroma Wheel. Because the reference standards are made with single chemicals, they can be prepared with great precision. However, they don't reproduce complex aromas like raspberry or asparagus, and they are not cheap and easy for the amateur beer enthusiast to make at home. An added frustration is that the descriptive terms for beer are confusing. In the "sulfury" category, for example, are "sulfury," "sulfitic," and "sulfidic," terms only a chemist could love.

THE APPEAL OF aroma wheels is that they organize product-specific smells into a few, easily recognized categories. As a result, food lobbies around the world have come up with their own versions. There is a chocolate aroma wheel from Switzerland and a Flavour Wheel for Maple Products, courtesy of Canada. There is a pan-European wheel for Unifloral Honey, and another for cheese (although with seventy-five different aromas it doesn't really simplify things for cheese fans). There is a South African brandy wheel, and the Berkeley-based perfumer Mandy Aftel has created a

Natural Perfume Wheel. Recently, some guys in the Philadelphia Water Department came up with a wheel for identifying the odors found in sewage. (Anyone who's lingered on the banks of the Schuylkill River knows that wastewater offers a particularly rich olfactory experience.) The world has gone wheel crazy, and we can expect to see more of them in the future.

The Perfumer's Problem

Odor space is an imaginary mathematical realm containing all possible odors. The aromas of wine and beer occupy only a small fraction of odor space—not nearly the full range of smells detectable by the human nose. Can smell classification work on a larger scale? Perfumes and colognes take up a bigger chunk of odor space: There are at least 1,000 currently on the market, with new ones added at a rate of about two hundred per year. Each has anywhere from 50 to 250 ingredients. If anyone has a lot of smells to keep track of, it is the perfumers who create them.

At its core, the practice of perfumery hasn't changed much since it came to full flower in Renaissance Italy. In those days there were no more than 200 commonly available ingredients, all derived from natural sources, either botanical (essential oils, gums, spices, and barks) or animal (musk and civet). By the late nineteenth century, discoveries in organic and synthetic chemistry created a host of new materials. Some were novel, man-made molecules; others were pure chemicals isolated from the complex mixtures found in nature. The result is that the modern perfumer's palette is far larger than his predecessor's. Learning these materials is a correspondingly bigger task. How does a perfumer keep it all straight?

The professional perfumers Robert Calkin and Stephan Jellinek explain: "The novice perfumer may well feel daunted by the hundreds of bottles containing strange and often unpleasant smelling materials that line the laboratory shelves. But for the talented student the task of learning to identify them is in fact less difficult than it may seem at first." The trick, according to these experts, lies in honing specific cognitive skills, namely learning new mental categories and how to fit new smells into them. To become a perfumer you don't learn to smell like one—you learn to think like one.

The first step in training is to learn the smell of the available ingredients. The leading teaching technique—the Givaudan method, created by the French perfumer Jean Carles—introduces students to the major ingredients using a matrix approach. Imagine a grid of rows and columns. Each row is a fragrance family: citrus, woody, spicy, and so on. Each column is a training session. In the first session, students smell column-wise one material from each family: lemon oil, sandalwood oil, and clove bud oil, for example. In the second session, the students smell new examples: bergamot oil, cedarwood oil, and cinnamon bark oil. This process continues for about nine lessons, by which time the students are familiar with the olfactory differences between families. Now comes the hard part—learning the "contrasts" within a family. Each subsequent session traverses one row of the matrix. In the citrus lesson, for example, students smell lemon, bergamot, tangerine, mandarin orange, blood orange, grapefruit, and lime. The goal, according to master perfumer and teacher René Morgenthaler, is for the student to create a personal impression of each ingredient. These individualized mental hooks are the key to remembering the fine discriminations needed to do perfumery. The graduate of nasal boot camp must recognize more than 100 natural materials and around 150 synthetics. The professional perfumer eventually becomes familiar with every material in his company's library—anywhere from 500 to 2,000 items—and is able to recognize every grade of each.

With the basic raw materials in mind, a trainee next learns to think like a perfumer. When a professional analyzes a fragrance or creates a new one, he does not think in terms of individual ingredients; he thinks of typical combinations called accords. An accord is a mixture of raw materials (rarely more than fifteen) that go together particularly well. Accords are the building blocks of perfumery. By combining several of them, the perfumer creates an initial sketch of the perfume, sometimes called a skeleton. In a way, creating a perfume is like writing software: a programmer starts with building-block software modules that already contain many lines of code. A computer program is built with many modules, just as a fragrance is assembled from accords. The analogy goes further— software is tested with iterative debugging; perfume is tested with repeated sniffing and tweaking of the formula.

An art form as subjective and personal as perfumery might be expected to resist computerization. In fact, the opposite is true. The practice of perfumery quickly adapted to the digital world in terms of tracking materials and recording formulas. At a more fundamental level, the perfumer and the software programmer share a similar mind-set that involves the logic of subprograms and modules. Some of the memory burden of those thousands of ingredients is relieved by computer technology. Perfumers browse the company's entire inventory of materials on-screen. They assemble a formula with a series of mouse clicks. They save everything: formulas, failed trials, and favorite accords. Software is an active partner in the creative process. It warns the user when two chemically incompatible materials have been selected, thereby avoiding a formula that discolors when exposed to sunlight. Most important, it continuously tallies the cost of the formula and displays it on-screen as dollars per pound of fragrance oil. No matter how great the creative latitude on a given project, a perfumer always works to a dollar limit.

Once a novice starts to think like a perfumer, he begins to develop a new way of smelling. Individual ingredients recede and whole fragrances emerge: he learns to smell the forest before the trees. Given a new men's cologne, he quickly recognizes it as, say, a Fougere type. Next he sniffs for the individual notes that define the

Fougere pattern: lavender, patchouli, oakmoss, and coumarin. After confirming these, he smells further, looking for a new twist or nuance that sets this formula apart from all the other Fougeres in the world.

Perfumers reduce the complexity of their world to a small, manageable number of fragrance families. They use well-known accords to simplify the process of scent creation. The perfumer's job is more about pattern recognition than about raw memorization; his mental map is uncluttered by free-floating details. Like most highly creative people, perfumers tend to be a little crazy; but they are not driven crazy from remembering thousands of smells.

The Shopper's Problem

Hundreds of perfumes are available for sniffing in department stores and boutiques. They range in style from restrained elegance to loud assertion, from distinctive originals to blatant knock-offs. How does one shop for scent amid this sensory overload? Lacking the perfumer's trained thought processes, the average person is completely at sea.

Fragrance houses—the companies that employ perfumers and create the juice for the Calvin Kleins and Cotys of the world—find it useful to organize perfumes by smell. The Haarmann & Reimer company published a fragrance genealogy that traces every style of perfume from its first appearance to the present day. It's a nasal Book of Genesis: In the beginning was *Jicky* (Guerlain, 1898), and *Jicky* begat *Emeraude* (Coty, 1921), and *Emeraude* begat *Shalimar* (Guerlain, 1925), and so on through *Obsession* (Calvin Klein, 1985) and those that followed it. (Although some ancestral scents are well-known classics, it is sobering to see all the brand names that meant so much in their time but so little today: *Moon Drops* (Revlon, 1970), *Touché* (Jovan, 1980), or *Aspen* (Quintessence, 1990).

Genealogies provide a sense of history, but they don't help one to shop in the here and now.

Another style of perfume guide lists each brand by fragrance family: florals, aldehydics, chypres, and so on. This isn't much help if you don't know what a chypre smells like. (The term covers a range of styles having in common a warm, woody character with an animal-like undertone.) If you like Estée Lauder's *Pleasures,* you can look up a dozen similar scents. What you won't find is a measure of how similar they smell. Nor will you find the exact ways they differ from *Pleasures*—are they stronger, spicier, greener, muskier?

Most people don't consult a reference book before shopping. They simply head for the department store. But once inside, things don't get easier. Each fragrance brand has its own counter, attended by its own salesperson who will show you only the perfumes she is paid to show. If your ideal scent is one counter away, it might as well be in a different universe. Sephora stores broke with retail tradition by introducing the "open sell." Brands are arranged alphabetically on the shelf, from Alan Cummings to Yves Saint Laurent. With no vested interest in any one brand, the Sephora staff is just as happy to sell you Alan as Yves. To introduce sensory logic to their store designs, the company has tried arranging perfumes by fragrance family: orientals here, florals over there. This may be the start of a badly needed rethinking of the retail experience.

Charts and guides, even those based on expert opinion, are still arbitrary views of odor space. They present the world according to one fragrance house, or more likely just its chief perfumer. No single classification has emerged thus far as the industry standard. If one did, it still wouldn't help the average shopper, because perfumers don't think like the rest of us. The professional detects rose de mai Bulgarian where the consumer smells flowers. The professional finds clear-cut differences among perfumes that strike most people as indistinguishably fruity-floral. What the average person needs is a map on which brands are arranged by how they smell to other average people.

PERFUME MAKERS speak to the consumer with two voices: Ingredient Voice and Imagery Voice. Here is a classic example of the Ingredient Voice, from a description of Estée Lauder's best-seller *Beautiful* (1985):

> Vibrantly feminine floralcy of rose, lily, tuberose, marigold, muguet, jasmine, ylang, cassis and carnation accented with fresh mandarin and bright fruity notes. Warm background accord of orris, sandalwood, vetiver, moss and amber.

Ingredient Voice assumes perfumer-level familiarity with more than a dozen raw materials, when in fact few civilians have ever smelled orris root or vetiver. Reciting a list of ingredients gives an illusion of precision. Even perfumers don't think of *Beautiful* as a list of ingredients; they might think of it as a big, complex floral type with an ambery warmth. Ingredient Voice doesn't help the casual shopper.

In contrast, Imagery Voice is all about atmospherics. The drama of seduction, passion, and mystery makes Imagery Voice the natural language of brand marketers and ad agencies. Listen to an actual vice president of marketing discuss a new men's cologne with a cosmetics industry trade magazine: "It's intended to target a young, stylish, hip, contemporary kind of guy." So far, so good. Aging, badly dressed nerds aren't known for buying a lot of cologne.

"The positioning of [the new brand] is really based all about capturing the pulse and energy of the city." Reasonable enough— discretionary consumer dollars don't chase listless, slow-moving, rural scents. But what does the new scent smell like? Hearken to the Imagery Voice:

> The fragrance notes themselves are city inspired, in that the top notes we describe as being powered by "living liquid air." It's

fused into a matrix of metal aldehydes and it captures the feeling of shiny steel and glass in a modern urban environment. It is very fresh and almost metallic on top; then it dries down to warmer, more sensual suede and woody notes on the bottom.

That's an impressive prose poem. Buried in it are two actual smells: suede and wood. Suede is fairly specific—one imagines the smell of a supple jacket or new pair of shoes. Woody, on the other hand, covers a lot of territory: pine, oak, cedarwood, redwood, cypress, and don't forget sandalwood. If Mr. Young Stylish and Hip wants to know what this new cologne smells like, he's just going to have to smell it for himself.

Imagery Voice combines ordinary adjectives (fresh, woody) with technical terms (aldehydes) and envelopes them in emotional verbiage ("the feeling of shiny steel"). The result is the marketing equivalent of a Jell-O mold at a church dinner.

NOTICEABLY ABSENT from the world of perfume is the identifiable voice of the independent critic. There is no Roger Ebert of scent. A scientific polymath and self-appointed expert named Luca Turin once tried his hand as a freelance perfume reviewer. Turin is serious about the aesthetics of fragrance and not a shill for any manufacturer. But his capsule reviews tend to be highly stylized. Here's a whiff: "*Après l'Ondée* evolves only slightly with time: its central white note, caressing and slightly venomous, like the odor of a peach stone, imposes itself immediately and retains its mystery forever." Turin makes *Après l'Ondée* sound both impossibly abstract and off-puttingly tactile. Meanwhile, a reader still doesn't know what it smells like.

Perfume wearers need a style of commentary that blends the aesthetic and the technical, like the road tests in *Car and Driver* that talk about sporty handling and trunk space in the same paragraph.

In 2006 the *New York Times* tapped Chandler Burr as its first-ever perfume critic. Burr rates perfumes with a conventional five-star scale and a writing style that tilts heavily toward the aesthetic: "This is the scent of the darkness that inhabits a Rubens, a warm, rich, purple blackness; *Pomegranate Noir* is like a box of truffles with the lid on, sweet bits of darkness, waiting." (OK, but what about the horsepower and mileage?)

Practical-minded perfume fans might prefer "Andrew," who pens cheeky analysis for the English newspaper *Metro*. Here's his take on *Live Luxe* by Jennifer Lopez: "It'd take a very brave/mad woman to wear this one. Ridiculously sweet and fruity, this is the fragrance equivalent of going out dressed as Carmen Miranda with a fruit cocktail poured down your cleavage. Invigorating but not for use in an enclosed space." Andrew recommends it "[f]or ladies who like to make an impression."

How come we have *Cigar Aficionado* and *Wine Spectator*, but no *Perfume Enthusiast*? This is a magazine publishing niche waiting to be filled. In the meantime, perfume bloggers are popping up all over the Internet: IndiePerfumes, Anya's Garden of Natural Perfumery, SmellyBlog, Scentzilla, to name just a few. As elsewhere in the blogosphere, this evolving community is a mixture of the personal and the professional, the serious and the whimsical. But the passion for fragrance is always there. These writers are pioneering new ways of describing scent. I think their efforts may produce a vibrant, robust, and very useful way of organizing the world of perfume.

The Big Enchilada

Perfumes, flowers, and wine occupy the sunny heights of the smellscape. Beyond lies the Dark Side, a swampland reeking of burnt rubber, rotten eggs, and the silent but deadly guy on the No. 33 bus. Few people aspire to study stench—there are no maestros of malodor. And yet, if we are truly to understand the sense of smell, we must account for the whole of it: the good, the bad, and the ugly. Where is the Universal Classification of Smell?

According to conventional wisdom, all major smell classifications can be traced back to the eighteenth-century Swedish naturalist Carl von Linné (1707–1778), known as Linnaeus. Linnaeus was the Big Daddy of scientific classification. In fact, he was a little obsessed with the topic: he classified plants and animals, rocks and sea creatures, and even his fellow scientists. Far from being a muddy-boot field biologist, Linnaeus was a bookish desk-jockey more concerned with defining the single "type" of a species than with the extent of natural variation. For this reason, some historians view him as a rigid essentialist who held back progress in the life sciences for decades. Still, his decision to assign a two-part Latin name to every species— something he regarded as a minor innovation—was a stroke of genius, and it became the basis of all modern taxonomy.

Linnaeus is widely credited among psychologists with inventing the first scientific classification of smells. Very few of them, however, seem to have read the actual treatise, published in 1752. Its Latin title, *Odores medicamentorum,* translates as "The smells of medicines," and this is the first big hint that Linnaeus's primary interest was not smells, but the medicinal properties of plants. He believed he could predict the therapeutic effect of a plant from its odor. To his way of thinking, nonsmelly plants were medically worthless, while strong-smelling ones had great pharmaceutical potency. Similarly, he believed sweet-smelling plants were wholesome, nauseous ones were poisonous, spicy ones were stimulating, and "noisome" ones were "stupefying." These effects were due to plant smells acting directly on human nerves. You can be forgiven if the views of Sweden's greatest scientist sound to you like those of a New Age aromatherapist in contemporary Santa Monica.

In grouping medically useful plants by odor, Linnaeus came up with seven classes that translate as fragrant, spicy, musky, garlicky, goaty, foul, and nauseating. His only concern was using smell to classify natural medicines; he did not intend to create a universal classification of all smells. In fact, he had little interest in smells *as* smells. (This explains the absence of such obvious odor categories as floral, fruity, woody, and leafy green.) Despite his focus on medical properties and his neglect of sensory qualities, European scientists viewed Linnaeus as the first scientific classifier of smells, and the results were a disaster—it sent smell researchers on a wild-goose chase that lasted for two centuries.

The next scientific smell classifier emerged toward the end of the nineteenth century. The Dutch physiologist Hendrik Zwaardemaker (1857–1930) was, by his own account, not particularly interested in smells. His lack of feeling for the topic shows in his work, where his main contribution was to add two new classes ("ethereal" and "empyreumatic") to those of Linnaeus and to create subdivisions within each class. The new version was complicated and made little sense as a comprehensive classification. (He was, after all, cramming every smell in the world into categories meant to organize only smelly medicinal plants.) Zwaardemaker labored to explain his system, but his tedious cross-referencing of previous classifications has all the prose sparkle of the IRS tax code. Like the system it expanded on, Zwaardemaker's classification was based entirely on one man's opinion, rather than on experimental data.

The German physiologist Hans Henning (1885–1946) relentlessly attacked the inconsistencies and absurdities in Zwaardemaker's classification. He took aim at Zwaardemaker's preference for lifting odor descriptions from novels and literary works rather than from the direct experience of his own nose. Henning insisted that sensory experience was superior to empty intellectualizing; his motto was "just smell it." His own classification, proposed in 1916, had two very important selling points: it was based on empirical data, and it came with a ready-made visual representation, the "odor prism." The image was a compelling one, orderly and neatly geometric. The six corners of the prism were each assigned a specific odor quality. Henning claimed that any odor could be located on the surface of the prism; its distance from any corner indicated the relative contribution of that odor quality.

Unfortunately, Henning overplayed his hand. The clean geometry of the odor prism proved irresistible to the scientific psychologists in America, who tested its feasibility in laboratories at Harvard, Clark, and Vassar. Initially enthusiastic, the Americans soon found his theory to be cumbersome and too vague to yield testable predictions. In their hands, it produced inconclusive results. Henning's initial theory was based on work with only a few experimental subjects; it now became clear that those subjects were extremely, if not suspiciously, consistent in their responses. (Wide person-to-person variability is a hallmark of odor perception; it's unlikely that randomly selected sniffers would agree as precisely as Henning's trio did.) In retrospect, there was always something too neat about Henning's idealized prism: its geometric elegance is undeniably appealing, but few areas of human experience are less linear than smell.

The dismantling of the odor prism by American psychologists ended the European tradition of armchair smell taxonomy. The search for a Universal Classification of Smell shifted entirely from philosophical reasoning to experimental research, and with it momentum crossed the Atlantic for good. Although as outmoded as the buggy whip, the odor prism persists in contemporary encyclopedias and textbooks, a testament to its iconic power.

IT WAS FRUSTRATION with Henning's prism that led the Americans Ernest Crocker and Lloyd Henderson—of the "10,000 odors" estimate—to invent a new system of smell classification. They began by selecting four "elementary odor sensations": fragrant, acid, burnt, and caprylic. Then they assembled a set of odors to serve as reference standards, by means of which any smell could be rated on a scale of 0 to 8 for each of the elementary sensations. Rose, for example, was rated 6 on fragrant, 4 on acid, 2 on burnt, and 3 on caprylic. Those four numbers (6423) became, presto change-o, a digital identifier for that particular smell. In the same way, vinegar was 3803 and freshly roasted coffee was 7683. A numerical specification of sensory quality is not that outlandish; the Pantone color standards, for example, use numbered samples to let graphic designers and printers communicate accurately.

The Crocker-Henderson system had wide appeal because it was based on empirical data and an open set of standards: anyone could play. Following its publication in 1927, the system was quickly commercialized; the complete set of reference odors could be ordered from Cargille Scientific, Inc., in New York City. It was soon being used by distillers, soap companies, the U.S. Army, and even the Department of Agriculture. Sensory psychologists initially gave the system positive reviews, but in 1949 researchers at Bucknell University dealt it a stunning blow. They found that untrained people couldn't sort the thirty-two reference odors into anything resembling the four elementary sensations postulated by Crocker and Henderson. Further, people were unable to arrange the eight odors within an elementary group in order of intensity. Because the Crocker-Henderson system was premised on elementary odors and intensity-graded smells within them, the new findings effectively undermined its logic. User enthusiasm vanished and the system eventually faded away.

ANOTHER BURST OF innovation in odor classification took place in the 1950s and 1960s when chemist John Amoore observed that people who were odor-blind to the stinky-feet smell of isovaleric acid were relatively insensitive to similar smells. He proposed that "sweaty" was a primary odor in the same way that red is a primary color. Amoore sought out molecules with similar shapes and smells that he thought might be the basis of other primary odors. (He eventually proposed seven of them: camphoraceous, musky, floral, pepperminty, ethereal, pungent, and putrid.) While he did succeed in finding other instances of selective odor blindness, Amoore's notion of primary odors did not hold up under rigorous sensory testing. In the end, the structural features of a molecule are not a reliable guide to the psychological realities of odor categories.

The latest attempts at odor classification use a technique called semantic profiling, an approach pioneered by the fragrance chemist Andrew Dravnieks in the 1960s, and still used today. Researchers hand people a long list of smell descriptors and have them check off as many as apply to a given odor sample. The hope is that with enough descriptors, smells, and statistical analysis, a pattern will emerge. And indeed patterns do appear—it is possible to point to groups of odors that share similar descriptions. The trouble is, this leaves us back where we started from—odors are described similarly because they smell similar. What we really want to know is, Why do they smell similar? For now, scientists are stumped—the molecular structure of odors isn't the answer, nor can we conjure categories from lists of adjectives. As a result, researchers today are reluctant to propose anything like the grand classifications of the past.

If HISTORY IS littered with the wrecks of Universal Classifications of Smell, we can still learn something from surveying the ruins. What they have in common is a surprisingly limited number of elementary categories: either 4, 6, 7, or 9, depending on who you like. The mind-boggling variety of smells in the world is reducible to a manageable handful of nameable odor classes, just as the brain carves the range of visible light into a handful of focal colors. Suppose one adopted the standard perfumery categories as an approximation of the pleasant sectors of smell space; this amounts to one or two dozen classes (woody, floral, fruity, citrus, etc.). What more would one need to encompass the stinks and stenches of the world? The fecal category would cover a lot of territory—from benign horse manure to the intolerable air in a rock concert privy. A category for urinous could include the sour smells in a nursing home and the heavier fourth-quarter reek of urinals at an NFL game. We'd have to add a class for retch-inducing smells—vomit and really stinky feet—and another for fishiness in all its gradations. Skunk, sulfur, and burning rubber could constitute yet another class. Finally, the putrid stench of rotting meat probably deserves its own banner. These six classes would capture most of the bad smells abroad in the world. Which is more amazing—the huge number of possible odors, or the tiny number of odor types?

CAN SUCH A stripped-down system of classification handle the olfactory complexities of the real world? It turns out that the human brain already does a pretty good job of reducing that complexity. The Australian psychologist David Laing was the first to tackle the relevant question: How many smells can we pick out of a complex mixture by nose alone? He began with a set of distinctive odors such as spearmint, almond, and clove, each easily identifiable on its own. He created mixtures—beginning with combinations of two odors at a time—and asked people to identify as many components as they could. The more odors he added to the mixture, the more difficult it became to identify even a single ingredient within it. The degree of difficulty was surprising. For example, in a mixture of three or more odors, fewer than 15 percent of people could identify even one component. Laing made the test easier: he gave people a target odor and asked them whether they could smell it in the mixture. Even then, they could rarely find the target in a mix of more than three odors. Could the problem be lack of skill? Laing tested perfumers and flavorists. The professionals were better than amateurs at identifying two and three items in a mixture, but even with their training and experience, they failed to pick more than three odors from the mix. Laing reasoned that mixtures of simple, single-chemical smells are somehow unnatural and hard to pick apart. So he repeated the experiments using as mixture components such complex odors as cheese and chocolate. The results were the same: no one could bust the four-odor limit. Were the individual smells not distinctive enough? Some odors, such as orange, almond, and cinnamon, blend together easily; perhaps those that blend poorly, such as mushroom, cut grass, and mandarin, are easier to pick out of a mixture. Laing found this was true to a point, yet the four-item barrier held firm.

Why are we so feeble at smelling our way through a bouquet? Our ability to gather olfactory information is formidable: the human nose detects single smells at extraordinarily low concentrations. We do a better job of collecting smells than we do of tracking them in a complex mixture. The Laing Limit suggests that the problem is not in the nose but in the brain. We have limited ability to think about smells analytically.

In the end, the question "How many smells are there?" may not be as relevant as "How many odor categories do we need to make sense of the world?" The answer to that question will reveal much more about how the brain handles the information that the nose provides.

CHAPTER 2

The Molecules That Matter

You cannot suppose that atoms of the same shape are entering our nostrils when stinking corpses are roasting as when the stage is freshly sprinkled with saffron of Cilicia and a nearby altar exhales the perfumes of the Orient.

—LUCRETIUS

STRICTLY SPEAKING, SMELLS EXIST ONLY IN OUR HEADS. Molecules exist in the air, but we can only register some of them as "smells." Odors are perceptions, not things in the world. The fact that a molecule of phenylethyl alcohol smells like rose is a function of our brain, not a property of the molecule. A tree burning in the forest does not smell if no one is there to smell it. The planet Mars has no atmosphere and is too cold for human life, yet the chemical composition of its surface suggests that if we could sniff it, it would reek of sulfur. Perhaps someday we will have the opportunity. Apollo moon-mission astronauts noticed that the lunar dust they tracked back into their craft smelled like wet ashes in a fireplace, or burned powder from a shotgun shell. Humans flying back from Mars may need to hang a little pine tree in the cockpit window.

Semantics aside, an odor perception is usually caused by a physical substance—molecules light enough to evaporate and be carried on air currents to our nose. (There are strange exceptions: some observers of the early aboveground nuclear bomb tests experienced a metallic smell within moments of the blast, and in the rare condition known as phantosmia, patients perceive a smell in the absence of any external stimulus.) The sensory cells in our nose convert a chemical signal (the molecule) into an electrical signal (a nerve impulse) that travels up the olfactory nerves to the brain for interpretation. Since airborne molecules trigger odor perceptions, we should, in principle, be able to match a molecule to every odor. Hydrogen sulfide smells like rotten eggs and amyl acetate smells like banana—how hard can it be to complete the list? Very hard, it turns out. Most aromas in nature are elaborate bouquets, mixtures of dozens if not hundreds of different molecules.

Prior to 1955, complete chemical analysis of the aroma from a cup of coffee was beyond the reach of routine science. It would have been taken years to extract, isolate, and purify the scores of volatile molecules found in it. The invention of gas chromatography in the mid-1950s made possible the rapid analysis of aromatic mixtures and revolutionized the science of smell. Despite its importance, the gas chromatograph (or GC) remains little known to an otherwise technology-savvy public. Take a smelly substance—an apple or an oyster, it doesn't matter—put it in a blender, then run it through a GC, and you will get a visual record of its volatile components.

At the heart of the GC is a Slinky-esque coil of very thin tubing that would stretch ten to thirty meters if unwound. As a first step, the smell sample is injected into the coil, where it is absorbed into a polymer that coats the inside of the tube. The Slinky sits in a little oven, which heats up in preprogrammed steps over the course of two minutes to two hours, depending on the setup. A stream of helium gas enters one end of the coil and exits the other. As the temperature rises, odor molecules are driven out of the polymer and into the gas stream. The process is orderly: each type of molecule evaporates and enters the stream at a specific temperature, depending on its molecular weight, and emerges from the end of the coil in a burst roughly two seconds long. The amount of material in each burst shows up as a peak on a timeline. The more molecules, the bigger the peak. A pure sample of a single chemical, say phenylethyl alcohol, yields a single peak. A complex mixture like rose oil produces a series of peaks, varying in height, representing the more and less plentiful components in the mixture.

Because it is highly detailed and unique to each sample, the visual profile created by the GC is often likened to a fingerprint. The difference is that a fingerprint is static—a direct physical impression—while the GC is dynamic: it takes a complex smell and pulls it apart in time. Perfumers liken a smell to a musical chord; if this is the case, then the GC plays it as an arpeggio.

As individual odors emerge from the GC, they can be fed into another device called a mass spectrometer, which provides a definitive identification of the molecule. By the mid-1970s the GC/MS linkage had been automated and labs around the world were churning out detailed chemical analyses of natural products. This was a mixed blessing for smell scientists. Run orange pulp through a GC/MS and you get a laundry list of volatile components. Do they all smell? Do they all contribute to the total orange aroma? How can we tell?

Since the early days of GC, chemists have sniffed at the exiting gas stream to see if they could recognize the emerging components by nose. Some volatiles, such as carbon monoxide, are entirely odorless to the human nose; otherwise each GC peak corresponds to a distinct smell. The size of the peak is not a reliable index of odor power. A big peak may deliver very little odor (which means the molecule is not very smelly) and a tiny peak may pack a punch (the molecule is a potent odorant). Cornell University chemist Terry Acree pioneered what is known as gas chromatography-olfactory or GC-O, which is essentially a formalized way of sniffing the GC vent to correlate smells with specific molecules. Acree devised a way to express numerically the relative odor potency of each chemical within a complex sample. He divides a chemical's concentration in the sample by the minimum concentration needed to smell it on its own. Molecules with an odor impact index hovering around 1.0 are just at the level of detectability. Molecules with high multiples contribute more to the overall odor, while those with multiples less than 1.0 are seldom detectible; at best they lend a grace note to the overall composition.

Hey Beavis, Pull My Finger

One might expect the chemistry of certain bathroom malodors to be well understood. What other stinks are experienced on so personal a basis? For years, medical students were taught that the main ingredients of fecal odor were skatole and indole, nasty-smelling molecules created by the breakdown of meat protein during digestion. This claim persisted in textbooks despite never having been confirmed by direct chemical analysis. The shit finally hit the gas chromatograph in 1984 when researchers in Salt Lake City ran some poop through a GC and sniffed the results. Skatole and indole, although present in the sample, contributed relatively little to the typical fecal odor. The key actors turned out to be sulfur-containing compounds such as methyl mercaptan, dimethyl disulfide, and dimethyl trisulfide. Despite this dramatic reversal of conventional medical wisdom, the gastroenterological community remained unmoved. Finally, in 1998, investigators at the Veterans Administration Hospital in Minneapolis took the next step and performed an exacting chemical and olfactory analysis of farts. Their experimental methods were straightforward: "To ensure flatus output, the diet of the subjects was usually supplemented with 200 g pinto beans on the night before and the morning of the study." Gas capture was simplicity itself, though the details are squirm-inducing: "Flatus was collected via a rectal tube . . . connected to a gas impermeable bag." When the bags of ass-gas were analyzed, the main contributors were once again sulfur-containing molecules: hydrogen sulfide, methyl mercaptan, and dimethyl sulfide.

By comparing bean-powered samples from men and women, the intrepid Minnesotans were able to settle a long-running dispute between the sexes. The data proved (as men have claimed for centuries) that the farts of women are stinkier, on a volume-for-volume basis, than those of men. Since men produce a greater volume than women, however, the overall gag factor remains about even. As part of their research, the team tested a device called the Toot Trapper, a fabric-covered foam cushion coated with activated charcoal. The cushion is worn inside one's pants and, according to the manufacturer, absorbs the offensive odor of intestinal gas. The Minneapolis team tailored a pair of fart-proof pants from Mylar sheets and duct tape. When volunteers wore the pants along with a Toot Trapper, the captured gas was indeed less smelly. ("Toot Trapper" strikes me as a lame brand name for this useful product. If I were the marketing consultant, I'd go with something more robust, like "Blast Master 3000.")

Lyrical accounts of child-rearing dwell on the wonderful smell of a baby's head. Less sentimental observers note that infants are prodigious gas-factories. In 2001 a group of pediatricians found that diet affects the chemical composition of baby farts (technically, they analyzed the gas produced by poop samples stored at body temperature for four hours). The gas from breast-fed babies was heavy on (odorless) hydrogen and very low on stinky methyl mercaptan. Babies fed milk-based formula had intermediate levels on every gas measured. Infants fed soy-based formula produced a lot of hydrogen sulfide (rotten-egg smell) and also the most methane. The good news is that methane is odorless; the bad news is that it contributes to global warming.

Another cherished belief is that one's own little bundle of joy produces better-smelling poop than the other kids. Remarkably, this belief holds up under strict scientific scrutiny. Mothers of fourteen-month-old babies contributed dirty diapers, which were sniffed from cardboard buckets. Each mother compared a diaper load from her kid to that of an anonymous sixteen-month-old who provided the reference sample. The other baby was stinkier when the dueling buckets were unlabeled; labeling the buckets (e.g., "Jason" versus "Other Baby") didn't increase the effect, nor did switching the labels reduce the effect; this means the mothers were not letting maternal pride interfere with their odor judgments—they really do find other children stinkier. This study also proves that some sensory psychologists have way too much time on their hands.

Reefer Madness

One complex botanical smell has had an outsized cultural impact on the nation. Rod Blagojevich captured it well when, during his campaign for governor of Illinois, he admitted that he had smoked marijuana, saying "it was a smell that we all, in our generation, are familiar with." He added, "I didn't like the smell of it." In contrast, Andy Warhol allegedly said, "I think pot should be legal. I don't smoke it, but I like the smell of it."

I once received a phone call from a graphic artist when I worked for the fragrance company Givaudan Roure. He was designing the booklet for a solo CD by a member of a well-known rock band and wanted to print it with ink that smelled like marijuana. Could my company supply such a smell? His request put me in an awkward spot. Technical hurdles were not the issue; our perfume chemist assured me that he could work up a good pot smell. (He also hinted broadly that the project would go faster if he had a high-quality sample to work from.) The decisive factor was financial—sales are measured in pounds of fragrance oil sold and by the price markup on the raw materials. In this case, the expected sales volume was minuscule and not worth the time perfumers would spend on it. Still, the project held a certain allure.

The more I thought about it, the more complications came to mind. Can one replicate the smell of pot without using delta 9-tetrahydrocannibinol (or THC), the psychoactive ingredient? If so, could it still get you busted by a drug-sniffing dog or your homeroom teacher? Would my company be legally liable for the consequences?

THC, and its chemical cousins, are not volatile and are therefore odorless. If a chemist stripped the THC from pot, the result would be genuine-smelling but buzz-less, the psychedelic equivalent of decaffeinated coffee.

When I reach him on the phone, I find Dr. W. James Woodford to be a genial fellow with a Southern accent. He is a fragrance and flavor chemist and the man who invented the first drug pseudoscent. Early in his career, during a stint as a guest researcher at England's New Scotland Yard, he encountered large samples of contraband cocaine. Woodford knew that the pure cocaine alkaloid was odorless, but when he sniffed it in the evidence room he noticed a distinct aroma. When exposed to air and moisture, cocaine chemically degrades and yields a sweet, prunelike odor. Woodford's scientific curiosity was piqued, and he traced the scent to a molecule called methyl benzoate. Methyl benzoate is found in flower scents; there's lots of it in snapdragons and petunias, and some in tuberose and ylang. Perfumers use it all the time, especially in fragrances of the *Peau d'Espagne* type.

Cocaine is illegal, as are its direct chemical precursors and metabolites. Woodford managed to replicate the scent of cocaine with methyl benzoate and a few other ingredients, all of which are chemically unrelated to cocaine and therefore perfectly legal. Woodford patented his drug pseudoscent formula in 1981, and the government was soon using it to train dogs and drug enforcement personnel. Woodford let the government use it for free. "I didn't make any money off of it," he says. Others were not so charitable, and soon an entire industry blossomed. The Sigma-Aldrich chemical supply company, for example, carries Sigma Pseudo™ Narcotic Scent Cocaine formulation, priced at $37.20 for 100 grams. They also sell an LSD formulation and another that mimics the scent of pot. Forensic chemists at Florida International University have created a fake Ecstasy aroma.

Drug dogs trained to find cocaine are, in fact, recognizing the scent of methyl benzoate rather than the cocaine molecule itself. This displacement effect is true for other major targets of drug dogs. Ecstasy gives itself away through the cherry-pie scent of piperonal, and methamphetamine has a characteristic cherry-almond scent from benzaldehyde. So yes, dogs find the drugs, but they should really be called drug-associated-odor-sniffing dogs.

Fragrance clients are nervous about having too real a pot smell for fear of alerting drug dogs and police. What if drug traffickers used these for their own ends? They could flood an airport with pseudoscents and sneak their contraband through while dogs and cops are chasing false leads. It hasn't happened yet, but Woodford recognizes the danger. "There's potential for mischief," he says.

Since THC itself is odorless, what gives pot its characteristic aroma? The natural product chemistry of marijuana is complex. Depending on the exact technique used—headspace capture to analyze the scent given off freely by the plant, or steam distillation to extract its essential oils by force—there are anywhere from eighteen to sixty-eight volatile chemicals. Most of these are familiar to plant chemists; they belong to a class of molecules known as terpenes, which occur in the floral scents and essential oils of many species. Examples are beta-myrcene and limonene, which are found in nutmeg, orange oil, and basil, as well as marijuana. Of course, not every volatile chemical has an odor, and even those that do may not be present in sufficient quantity to be detected by the human nose.

To construct the definitive chemical profile for pot aroma, one needs to perform a GC-O analysis. Remarkably, no such analysis has been published, and we don't know for sure which molecules are critically responsible for its characteristic odor. Sensory analysis by enthusiastic amateurs can be found on the Internet and suggests a winelike variety of nuances. The authoritative-sounding Standard Smoke Report asks aficionados to describe fresh buds and smoke with terms that include ammonia, earthy, licorice, and peach. An evocative, if tongue-in-cheek, review of a Beck concert in Costa Mesa noted several varieties present in the haze above the Pacific Amphitheater: "the gorgeous and unmistakable aromatics of California Indica—a fine blend of orange-flavored Californian strains, sweet acidity and a delicate finish . . ." Can the average pothead really sniff the diff between California Indica and Super Afghani? I close my eyes and recall Bob Marley, barely visible in a wall-to-wall ganja cloud at the Santa Cruz Civic Center, and Jerry Garcia spotlit in the doobie fog of the Winterland arena. Distinct varietal aromas? Not back then. But the market has evolved. Jim Woodford, the forensic drug sniffer, tells me the East Coast product often smells "like minty oregano," while the West Coast version is generally "skunky."

THE INDEPENDENT perfumer Harris Jones once formulated a pot smell for a client who manufactured scented candles. He included beta-pinene and limonene and all the rest, but to achieve a realistic final impression (or, as he puts it, a good "touch"), Jones found he needed a skunky note. He did some research on skunk secretions and concocted his own Pepé Le Pew formula. He prepared a solution of it at .01 percent, and used that at .5 percent in the total formula. The client loved it, but Jones ultimately dropped the project once he realized how many different ways he could be sued should drug dogs and angry parents find his rendition too accurate.

The realism provided by the trace of skunk may explain an anomalous finding in the 1970s by some Canadian psychologists. Exploring aversive odor conditioning as a way to interfere with marijuana intoxication, they put finely chopped strands of human hair into a joint; when lit, it produced a highly unpleasant smell. They gave doctored spliffs to volunteers who were already high from a smoking session in the laboratory. Contrary to expectations, smoking the stinky weed significantly increased the perceived high of the volunteers. Not only did the smell of burning hair fail to kill the buzz, it boosted it.

THE SWEET, FUNKY smell of pot is saturated with social attitude, as is patchouli oil, its counterculture twin, once used by hippies to mask the smell of pot. While patchouli has become a popular fragrance ingredient in consumer products, potlike notes rarely appear in the marketplace. Is it time for marijuana to become a brand-associated scent?

How well do purported pot re-creations measure up? A car air freshener in the shape of a cannabis leaf smells like rancid compost. The Showtime network ran a scented ad in *Rolling Stone* for the 2006 season of *Weeds*. The scent was as cheesy as the "Catch the Buzz" tag line: a blend of lawnmower clippings, potting soil, and cedar shavings (the poor man's patchouli). Cable industry pundits were coy, calling it "a distinctive herby aroma" evocative of "a certain something." Then there is *Cannabis Santal Eau de Parfum* from Fresh, a fragrance division of France's LVMH. "A forbidden blend of patchouli, cannabis and rose, this sensual fragrance captures the raw energy of a man and the desire for him." I stopped by the ultra-cool Fresh boutique on Spring Street in lower Manhattan to give it a try. The Fabio lookalike at the counter recited the ingredients with impressive accuracy, but unfortunately he had been trained to spray the wrong end of the perfume blotter. (For future reference, dude, you hold the wide end and spray the narrow end.) *Cannabis Santal* was pleasant, with a nice patchouli note, but it didn't come within a bong's length of smelling like real pot. In the end, the most these commercial promotions dare to do is wink at the consumer and say "made ya smell!"

Smellscape in a Bottle

John Muir experienced an olfactory epiphany on the upper reaches of California's Feather River. For a few minutes the smellscape of the Sierra foothills revealed itself to him in all its swirling splendor.

> The air was steaming with fragrance, not rising and wafting past in separate masses, but equally diffused throughout all the wind. Pine woods are at all times fragrant, but most in spring when putting out their tassels, and in warm weather when their gums and balsams are softened by the sun. The wind was now chafing their needles, and the warm rain was steeping them. Monardella grows here in large beds, in sunny openings among the pines; and there is plenty of bog in the dells, and manzanita on the hill-sides; and the rosy fragrant-leaved *chamaebatia* carpets the ground almost everywhere. These with the gums and balsams of the evergreens formed the chief local fragrance-fountains within reach of the wind.

Muir's image of fragrance fountains is wonderful, yet his dry references to Latin scientific names make us thirst for a more compelling description. Take the large beds of *Monardella*, for example; what does it smell like? *Monardella* belongs to the mint family. Given the locale, Muir was probably describing coyote mint

(*M. villosa*) or pennyroyal (*M. odoratissima*). I have hiked through ground covered in California pennyroyal and inhaled the fresh fragrance rising from the bruised leaves beneath my boots. Muir's description of "rosy fragrant-leaved *Chamaebatia*" leads one to imagine a pleasing floral scent, but this couldn't be further from the truth. *Chamaebatia* is a member of the rose family, but the plant he smelled (*C. foliolosa*) is unique to California. Its leaves are dull green feathery fronds, sticky with resin, and it sports tiny white flowers. The Miwok Indians called it *kit-kit-dizze*, but the settlers knew it as "Sierra Mountain Misery" or "Bearclover," names that reflect its pervasive, heavy aroma, akin to cooked artichokes or dilute cat urine. On a hot day in the Sierra Nevada, this musty smell rises like a tide and covers the land all the way from the Feather River where Muir inhaled it, past Lake Tahoe, down to Yosemite and to the southern foothills in Tulare County.

If only Muir's prose were as aromatic as his visual images. He rouses our curiosity, but can't sate it. We want to sniff. We want to dip a cup in the showering fragrance fountain. Why can't someone give us Muir's afternoon on the Feather River?

For nearly all of human history, capturing a scent from nature meant collection and extraction. Heaps of flower petals and baskets of resin were gathered and their essence stripped out with heat or solvents. The result of this harsh processing may be beautiful on its own, but it is a distorted and distant version of the original. The recent quiet revolution in technical chemistry has changed the way we capture scent, in addition to helping us understand its components. By the mid-1970s, GC/MS technique had become so sensitive that the amount of sample required for analysis was 10 to 50 micrograms, an incredibly small quantity. As my former colleague the Swiss fragrance chemist Roman Kaiser describes it, this is "approximately the amount given off by a moderately fragrant flower over the course of one hour." Kaiser and a few other experts developed nondestructive means of collecting scent. They take it from the air (or "headspace") surrounding the sample. Whether it's an orchid on the vine or a fruit on the branch, they don't physically disturb the odor source; they merely place a glass bulb around it and use an electric pump to suck the headspace through a molecular "trap"—a tube full of porous polymer that absorbs the scent. The trap can be stored and its captured scent later injected into a GC/MS back in the lab.

Headspace capture lets us analyze smells as they are produced in nature and as they are perceived by their intended audience (usually bees, bats, and butterflies). By analyzing the composition of a flower's living scent, rather than the oils extracted from its crushed petals, perfumers can better mimic the real thing back in the studio. The rarest specimens, unavailable in sufficient quantities for traditional extraction, can now be studied. (Having pioneering headspace analysis, Kaiser now uses it to study and preserve the scent of rainforest species threatened with extinction.) Other possibilities abound: the scent of a ripening strawberry can be traced as it changes hour by hour, as can the fragrance of a night-blooming desert flower as it varies from dusk to dawn.

I decided to enlist Roman Kaiser's help in tracking down the smelly essence of Sierra Mountain Misery. At the end of a camping trip in July 2006, I collected a few sprigs of it on the roadside a few miles west of Sonora Pass. I zipped it into a sandwich bag and stashed it in the beer cooler to keep it fresh as I drove down to Berkeley. I made it to Kinko's just before the express shipping deadline. Waiting in line, I spotted a sign listing restrictions on international shipments; among the forbidden items was plant material. Damn it. How would I get this stuff to Roman while it was still fresh? I stepped up to the counter, placed my bag of suspicious, leafy green plant material on it, and took a deep breath. "I'd like to send this express to Switzerland." "And what is it you're shipping?" asked the clerk. I flushed. "It's a . . . scientific sample." The manager looked over the top of his glasses; was he giving me the hairy eyeball? "Well, then, you'll need to fill out this international label." Phew. Good old Berkeley.

Three days later the Mountain Misery was stinking up Roman's lab in Dübendorf, a village outside of Zürich. Working his usual magic on the GC, he soon extracted around four dozen molecules: a smelly stew of terpenes such as alpha-pinene, beta-pinene, and plenty more. Most of them he could find in his extensive database of fragrant molecules; a few others would need months of work to identify fully with mass spectroscopy. Happily, this wasn't necessary to pin down the source of the distinctive smell; none of the mystery molecules smelled like Mountain Misery. Remarkably, one molecule, present in only trace amounts (.01 percent), was responsible for 95 percent or more of the cooked-artichoke smell. The chemical at the heart of John Muir's Sierra smellscape turns out to be 1-hexen-3-one.

Hexenone has been fingered as a key aroma in aged milk, cream, and butter; it also has a starring role in linden honey and fresh raspberries. This illustrates that a complex stew of volatile molecules can smell a lot simpler than what is implied by its lengthy ingredient list; one chemical can dominate an entire bouquet. Another lesson: abundance is not a reliable clue to odor impact; in this case, a single molecule from a single plant provides the aromatic background for an entire ecosystem And finally, it shows that a talented fragrance chemist can find the single molecule responsible for John Muir's poetic impressions of the Sierra Nevada foothills.

FROM THE WHISPERED fragrance of a single exotic blossom, it is but a step to capturing an entire smellscape. No one had a surer grasp of the grand scale of the American smellscape than Walt Whitman.

The conceits of the poets of other lands I'd bring thee not,
Nor the compliments that have served their turn so long,
Nor rhyme, nor the classics, nor perfume of foreign court or
*   indoor library;*
But an odor I'd bring as from forests of pine in Maine, or
*   breath of an Illinois prairie,*
With open airs of Virginia or Georgia or Tennessee, or
*   from Texas uplands, or Florida's glades . . .*

—WALT WHITMAN, *Leaves of Grass*

Pine forest or prairie, seashore or bayou, the essence of the ambience is there for the taking. To carry it away, all you need is a pump and a trap. To reproduce it is not trivial—it's a matter of money and determination—but it is firmly within our technological grasp. We can re-create the scent of coyote mint and pennyroyal and Sierra Mountain Misery. We can project them into your living room or office cubicle. Imagine them unspooling in slow transitions—a diorama for the nostrils—while you listen to Muir's afternoon on the Feather River, or to Whitman's ode to the American outdoors. What would you like to smell? For myself, I'd vote for the sea breeze at Point Reyes and the scent of the redwoods at Big Sur.

IN HIS 1947 MEMOIR *Speak, Memory*, the novelist and butterfly expert Vladimir Nabokov recounts a moment from one of his summertime collecting trips:

Unmindful of the mosquitoes that coated my forearms and neck, I stooped with a grunt of delight to snuff out the life of some silver-studded lepidopteran throbbing in the folds of my net. Through the smells of the bog, I caught the subtle perfume of butterfly wings on my fingers, a perfume which varies with the species—vanilla, or lemon, or musk, or a musty, sweetish odor difficult to define.

Scented butterflies are not exotic or rare. The Green-veined White, for example, is common throughout Europe and in parts of the United States, where we call it the Mustard White. To the British lepidopterist George Longstaff, its "strong and distinct" odor resembled lemon verbena. Back in 1912, he wrote: "It is curious that to this day so few persons are practically acquainted with the scent of the Green-veined White. When, at the Brussels Conference, in 1910, I caught a male *G. napi* in the beautiful garden of the Congo Museum, and demonstrated the scent to half a dozen entomologists present, none of these gentlemen had perceived the scent before, though at least one of them was a very eminent observer." The situation hasn't changed much in the last hundred years. No current field guides mention the scent of the Green-veined White—or of any species, for that matter. The fussy "butterflies through binoculars" crowd discourages physical contact with actual insects, but there are plenty of Mustard Whites in the Rocky Mountains, and you don't have to be as brutal as Nabokov. Go ahead and catch one for yourself. Sniff and release.

In Longstaff's field notes, one finds an astonishing range of butterfly odors. Some are like confections (vanilla, chocolate, burnt sugar), others like flowers (freesia, jasmine, heliotrope, mango flower, honeysuckle, sweetbriar). Yet others are like herbs and spices (cinnamon, lemon verbena, orris root, sandalwood, musk). Longstaff also found a spectrum of unpleasant scents, some reminiscent of cockroach or muskrat, others of rancid butter, butyric acid, vinegar, acetylene, musty straw, cow dung, horse stable, horse urine, and ammonia.

We now know that the lemony body odor of the Green-veined White contains alpha-pinene, beta-pinene, myrcene, limonene, linalool, p-cymene, neral, and citral. (The first five ingredients are also found in cannabis oil. Why should a psychoactive hemp plant and a butterfly share odors? Nature is wonderfully strange.) Males of the Green-veined White have another scent, which they hold in reserve for special occasions. It is methyl salicylate, easily recognized as the odor of wintergreen (or Pepto-Bismol). The male uses it as an antiaphrodisiac: he transfers the scent into the female at mating and it discourages other males from copulating with her afterward. Related species have their own versions of this turn-off tactic: the Small White uses a blend of methyl salicylate and indole; the Large White uses benzyl cyanide. These chemical counter-measures can backfire, as when the Large White's antiaphrodisiac aroma draws unwelcome attention of a tiny parasitic wasp called *Trichogramma brassicae*. When a female wasp smells a recently mated Large White female, she grabs on and hitches a ride. As the butterfly lays her eggs, the wasp parasitizes them by laying her own eggs inside them, and her young later use the butterfly's eggs for food. In the end, the male Large White who tried to defend his genetic investment ended up sacrificing some of his potential off-spring.

NATURAL BOTANICAL scents have a soft-focus, flower-child ambience about them. They are perceived as innocuous and innocent, a gift from Earth Mother Gaia to aromatherapists everywhere. In reality, they are biological communication systems, a way for plants and animals to talk to each other. This also makes them instruments of deception and treachery. Once a smell is used as a signal, other organisms can turn it to their selfish advantage. (Ask a female Large White how she feels about the parasitic wasp on her back.) A Mediterranean plant called the dead-horse arum fakes the stench of rotting meat. It attracts blowflies looking to lay eggs on a nice ripe carcass. The blowfly gets fooled into pollinating the plant for free, traveling from one stinky plant to the next carrying pollen on its legs, in what has been called "a striking example of evolutionary cunning that exploits insects for pollination purposes." Other examples are more sinister and almost perverse. An Australian orchid emits a smelly molecule called 2-ethyl-5-propylcyclohexan-1,3-dione, which happens to be the exact molecule produced as a sex attractant by females of the wasp species *Neozeleboria cryptoides*. When the orchid joins the action, the result is an aroma-based, cross-species sexual deception in which hapless male wasps attempt to copulate with the orchids. In the end, the orchid is pollinated and the male wasp is frustrated. Sex and exploitation are never far apart.

In nature, smells also serve defensive purposes. Essential oils, cherished as healing elixirs by aromatherapists, are really weapons in the ongoing battle between a plant and its predators. Take the orange tree as an example. It provides three different materials used in perfumery: neroli oil from its flower, orange peel oil from its fruit, and pettigrain from its leaves. Orange trees didn't evolve for the perfumer's convenience. Flowers smell good to attract pollinators; fruits smell and taste good to attract seed dispersers. A leaf releases volatile aromatic compounds as soon as an herbivore bites into it. This makes the leaf unpalatable or even toxic to the attacker (a caterpillar, say) and simultaneously alerts predators (such as wasps) that food is available. To an aromatherapist the orange tree is a repository of healing oils; to a caterpillar it looks like a weapons depot ringed with alarms and booby traps.

In college I lived for a time near the eucalyptus grove at the West Gate of the Berkeley campus. I loved to walk through its aromatic shade on the way home from class. The fresh astringency of the trees, like the fog that sometimes shrouded them, was to me a key element of Bay Area aesthetics. Back then I took a simple pleasure in that smellscape, and I still do. But today I also see it another way: as the lingering haze of biological warfare. Eucalyptol, chief among the fragrant molecules wafting about the West Gate, wards off leaf-eating bugs and suppresses the growth of seedlings of competing tree species.

The Web of Nature

Near Guaraqueçaba in southern Brazil is a remnant of the rain forest that until recently covered all 4,650 miles of the country's Atlantic coast. While prospecting there one spring for unusual smells, Roman Kaiser found the forest suffused with a strong fruity-floral scent. He tracked it to a tree with white bottle-brush flowers. Nearer the tree the scent took on a blackcurrant quality; close to the flower itself the smell resembled cat pee. With chemical analysis, Kaiser was able to trace both odors back to a single molecule: 4-mercapto-4-methylpentan-2-one, or MMP. (It is one of many molecules whose odor character depends on airborne concentration.) For most people, that would be the end of it: Odd Molecule Found in Exotic Locale. But Kaiser—a man with a chemist's brain and a perfumer's heart—has probably sniffed more GC samples than any living human. For him, MMP isn't a singularity—it's one node on a web of connections. Follow this molecule through the web and you'll find yourself transported all over the world. MMP is a key aroma in Japanese green tea, grapefruit, basil leaves, tomato leaves, box tree, cabernet sauvignon wine, and *Paeonia lutea* (the Tibetan peony). Is this a fluke? Or is 4-mercapto-4-methylpentan-2-one the clue to a hidden pattern in nature?

Since the advent of GC-O studies in the 1980s, chemists have analyzed everything from tomato paste to parsley, boiled beef to baby farts. In each substance they find many volatile molecules, yet only a few that are responsible for its characteristic aroma. Scientific journals are loaded with such studies, which are all cross-referenced in print. Imagine that this information is digitally organized and can be accessed as coolly and smoothly as Chloe calls up building diagrams for Jack Bauer on *24*. Each natural substance has its own web page listing key odorants—one can hyperlink from molecule to substance in any direction. Start, for example, with the home page for fresh oysters from the coast of Brittany. They contain 1-octen-3-one, which produces a mushroomy citrus note fancied by oyster lovers. Click on 1-octen-3-one, and you find yourself on the home page for Moroccan sardines, which they express this molecule after sitting on ice for a couple of days. In browsing the sardine page you find that fresh ones have a pleasant seaweedy scent traceable in part to (E,Z)-2,6-nonadienal. Click on that molecule and you are returned to the Brittany oyster home page. Why? Because (E,Z)-2,6-nonadienal is a characteristic odor molecule in fresh oysters.

Let's play the game again, this time starting with dimethyl sulfide, another key oyster odorant. It shows up in tomato paste, spoiled refrigerated chicken, and pinto-bean farts. Jump to the spoiled chicken page and click on methyl mercaptan; this will take you back to farts, or on to feces and french fries. From feces we can transfer to dimethyl trisulfide, which leads to Asian fish sauces and Gewürztraminer wine. Another key to the varietal character of Gewürztraminer is *cis*-rose oxide. Follow the link to *cis*-rose oxide and you see that this molecule is also responsible for the floral quality of fresh lychee fruit. On the lychee fruit home page you find that another potent odor is 1-octen-3-ol; clicking on it takes you to the Brittany oyster home page. Why? Because 1-octen-3-ol lends an earthy odor to both French oysters and lychee.

Is there a profound meaning in the hyperlink path from oysters to spoiled chicken to feces to Gewürztraminer to lychee and back? I doubt it. It's just Six Degrees of Kevin Bacon played with molecules. The olfactory web of 4-mercapto-4-methylpentan-2-one that links green tea to peony is not unusual. A given odor molecule turns up time and again; nature is economical and uses the same molecule different ways in different organisms.

⌒

BY 1974, ROUGHLY 2,600 volatiles had been identified in food. By 1997 the estimate had swelled to 8,000 and was predicted to climb eventually to 10,000. These are large numbers. They would be even larger if we included volatiles from nonfood items like airplane glue, dirty socks, and that crust of dried vomit under the backseat of the family minivan. Add them all up and the numbers are overwhelming. When it comes to potential smells, nature's bounty seems infinite.

What does all this molecular variety mean for the sense of smell? If the same chemicals turn up repeatedly as key smell ingredients, what impact does the rest of nature's chemical diversity have on the human nose? One answer is that we are missing most of it: we read the olfactory headlines and ignore the fine print. The field of sensory analysis confirms that only a fraction of the chemicals entering our noses from a given source make a difference to our perception of its odor. In most foods, for example, only a few of the volatiles detected by chemical analysis are present at nose-perceptible concentrations. Of the 400 or more volatiles found in a tomato, for example, only sixteen reach the threshold of human perception. One expert figures that fewer than 5 percent of the volatiles in a food actually contribute to its aroma. Perhaps odorants aren't as numerous as they seem.

So-called aroma models take this insight even further. To create an aroma model for french fries, for example, scientists run a batch through the GC/MS and generate a complete list of all the volatiles. Their goal is to create a fully realistic french-fry aroma using as few of the volatiles as possible. They begin by selecting odorants present at concentrations well above our sensory threshold. If a blend of those doesn't match the original aroma, they extend the list to include odorants at or below the sensory threshold. Once a blend closely matches the full aroma, it is tested further. One by one, odorants are subtracted from the formula. If the resulting formula smells less realistic, the subtracted odorant is restored. If the subtraction makes no difference, that odorant is dropped. The final aroma model is one of irreducible simplicity—a stripped-down formula that smells complete to the nose. An authentic french-fry smell, for example, can be made from nineteen ingredients. This includes a trace of stinky methyl mercaptan—without it, the formula lacks the necessary boiled-potato character.

Aroma models have been developed for Swiss cheese, Camembert, basil, olive oil, and baguette crust, among other things. These whittled-down formulas all point to the same conclusion—most volatiles in a food add nothing to its smell. A high-fidelity odor replica can be created from one or two dozen ingredients. A classic example is the cup of coffee. Chemists have been analyzing coffee aroma for more than 100 years and have found more than 800 different molecules. Using aroma models, German scientists found a mere twenty-seven high-impact molecules in medium-roasted Arabica coffee; they made a high-fidelity model using only sixteen of them.

The sensory logic of aroma models can be extended to nonfood areas, and may even have applications for environmental issues. Livestock feeding operations, for example, generate a big, messy stink that can annoy nearby residents. A typical Iowa swine barn contains more than 300 different volatiles, which sounds like a lot of bad news for the downwind neighbors. Yet a recent study found that four molecules account for about 85 percent of the piggy odor. One of these—para cresol—has a smell that by itself closely resembles the overall barnyard odor. This discovery may turn an overwhelming odor problem into a manageable project. Instead of going after all 300 suspect chemicals in the swine barn, one might suppress a handful of character-defining molecules. Pinpoint sensory targeting could produce bigger benefits at less cost.

THE SUCCESS OF aroma models—those minimalist imposters—casts nature's abundance in a new light. Lifelike smells can be made from a handful of molecules, and the same molecules turn up in smell after smell. Is nature's chemical cornucopia really so impressive if only a tiny portion of it matters? And what does it say about our sensory abilities if there is so much more out there than meets the nose?

Terry Acree, the Cornell scientist who helped develop GC-olfactometry, has the numbers to back this up. He searched through hundreds of food-aroma studies and made a list of volatiles present at smellable concentrations. The first edition of the FlavorNet list was posted online in 1997. It contained three hundred chemicals. Today he has posted about eight hundred. Acree expects the list to top out at fewer than one thousand. In other words, all the smells in nature are built from fewer than one thousand smellable chemicals. What are those other thousands of volatiles doing? They may subtly round out a scent, give it shading and complexity. Acree speculates that many of them are intended for the noses of creatures other than ourselves; the scents of nature are largely a chemical conversation between plants and animals, and humans merely eavesdrop. Just as we are blind to certain patterns on the wing of a butterfly or the petals of a flower because we cannot see in the ultraviolet, our mammalian noses are not tuned to certain olfactory broadcasts.

It is odd to think that a childhood's worth of olfactory memories can be boiled down to a pocket chemistry set. Were the tomato fields of Davis and the cooking ketchup of the Hunt's cannery—so powerfully evocative to me as a kid—just a particular shuffle of sixteen key odorants? Evidently so. Knowledge also brings insight. I now understand in molecular detail why my grade-school field trip to the Spreckels Sugar Company plant was such a stunning disappointment. As beets are processed into pure white refined sugar they first release geosmin (damp earth) and dimethyl disulfide (onion, cabbage, putrid). Later comes propionic acid (the pungent, rancid note in Swiss cheese and sweat), and finally hexenoic acid (musty, fatty). Those four notes were the heavy stew that oppressed my soul that day in third grade. Somehow, knowing that makes me feel better.

CHAPTER 3

Freaks, Geeks, and Prodigies

Don Giovanni: *Zitto: mi pare sentir odor di femmina!*
    [Hush! I think I scent a woman!]
Leporello: *Cospetto! Che odorato perfetto!*
    [My, my! What a nose!]
Don Giovanni: *All'aria mi par bella.*
    [And a pretty one at that.]

—Mozart, *Don Giovanni*

Take a few dozen people at random, and you will find a range of olfactory talent that stretches from *American Idol*–tryout bad to unbelievably excellent. There are people who cruise untroubled past the fetid plumes of garbage cans and subway vents, and others for whom the faintest milk fart escaping from an elderly relative is a nasal crisis. Olfactory sensitivity (technically, the lowest concentration at which someone detects a smell) is just one dimension of smell talent; other factors include an awareness of smells, and the ability to identify them and discriminate among them. Extreme variability is a hallmark of odor perception, and sensory scientists have identified many factors that contribute to it. It is now possible to answer a fundamental question: Who has a good nose and who doesn't?

⌒

THE FIRST THING to note is that people are not accurate judges of their own ability. When we asked people taking the National Geographic Smell Survey to rate their own sense of smell, we found a Lake Wobegon effect: most people were above average. The only way to assess someone's ability impartially is with a smell test. These come in two types: identification tests and threshold detection tests. The former ask you to put names to odors, the latter ask you to detect progressively lower concentrations of a smell. Smell tests have been commercially available for years, but were formally recognized as a medical device by the FDA only in 2006; this may explain why they are an underutilized part of the physician's diagnostic arsenal. The tests range from one-shot sniff tests, appropriate for quick screening during an office exam, to elaborate, hours-long testing with scores of odors that takes place in a research lab. Normal smelling is generally defined as a certain proportion of correctly identified odor samples, or a specific, very dilute concentration at which an odor should be smellable. An odd feature of smell tests is that the best one can score on them is "normal"; there is no test that rates levels of excellence, no equivalent to a 150 IQ. In fact, there is not even an official medical term for smell genius.

Because smell tests are designed to identify people with dysfunctional noses, they are finely calibrated for degrees of underachievement. At the lowest end of the scale are people unable to smell anything at all; they suffer from anosmia, the technical term for complete smell loss. One notch up the scale are people with hyposmia, which is the olfactory equivalent of being hard of hearing; like deafness, it can be mild or severe. It has been estimated that 1 to 2 percent of the U.S. population suffers from anosmia or hyposmia. In both cases, the most common cause by far is infectious disease. Severe colds, flu, and sinus infections inflame the tissues lining the nose and kill off sensory nerve cells. In severe cases, or after a lifetime of accumulated damage, areas that were once rich with nerve cells are replaced with nonsensory mucus membrane, and the tissue takes on a moth-eaten appearance.

Head injury is the second leading cause of smell loss. A blow to the head can sever some of the olfactory nerve fibers that travel to the brain through tiny holes in the base of the skull, at a location between the ears and behind the eyes. There's an old (and possibly true) story about a waiter carrying a tray of food at head height. As he exits the kitchen, the swinging door slams the tray into his forehead. Being a professional, he maintains his balance and proceeds into the dining room. As he serves the dishes, he realizes he can't smell a single one. The speed with which the waiter discovers his loss might be unusual—most people don't notice for days or weeks—but the mildness of the damaging blow is not. It takes very little force to cause smell loss. I cringe when I see kids heading the ball in a soccer game. I wouldn't bet on them becoming chefs or perfumers.

With the exception of a stuffy nose, smell loss is a long-term condition. Smell may return after a flu or sinus infection, as the damaged sensory cells are gradually replaced by new ones, but recovery can take months and your abilities may never return to their original levels. The probability of recovery declines with age. In cases of head trauma, the prospects are bleak; the severed nerve fibers rarely reconnect. Consider the results of a typical study: a year after their initial visit to the doctor, 32 percent of postinfection patients showed improvement, compared with only 10 percent of the post-injury group. The realization that millions of Americans suffer from smell loss spurred the National Institutes of Health to underwrite basic research into odor perception. The ultimate goal of this work was to find ways to cure smell loss. Yet, despite decades of substantial funding, effective medical treatment remains elusive.

Sudden smell loss is psychologically devastating. By far the biggest impact is on eating: anosmia steals the pleasures of the table. Without its aroma, food in the mouth becomes a bland, chewy mass, and drinks become equally flavorless. Faced with dull food, some people lose appetite, eat less, and lose weight; others eat to feel full and end up gaining weight. Smell loss can alter mood—patients often show symptoms of depression, and psychological well-being, friendship, emotional stability, and leisure activities all take a hit. Some people find that their sex life suffers. In the wake of smell loss comes the anxiety of constant vigilance. Anosmics worry about gas leaks, undetected fire, spoiled food, and lapses in personal hygiene. They adopt coping strategies such as frequent bathing and laundering. Anosmics report smell-related hazardous events—burning a pot or eating spoiled food—more often than normal smellers, but there is little data to suggest a higher rate of actual injury.

In rare instances, people are born without a sense of smell. As it's hard to miss what you've never known, people who are anosmic from birth tend to regard their condition with bemusement. A few even manage to find a silver lining. The ex-boyfriend of a young English anosmic told her, "You were the best girlfriend in the world. You let me bring curry home from the pub every night, and I could fart as much as I liked." One newspaper reporter who is smell-blind from birth regularly covers smelly stories for a major U.S. daily. This is either a heartwarming story of a disability overcome, or journalistic malpractice of the first order. Perhaps, in a zany way, it is both.

Somewhere off the main continuum of normal to partial to complete smell loss lie the bizarre pathologies of odor perception. A person with phantosmia, for example, perceives a smell when none is there. These olfactory hallucinations can be vague ("a chemical smell") or quite specific (one patient said, "It reminds me of a flower I smelled in Samoa"). Phantosmia is a tricky diagnosis for a doctor to make: the hallucinated smell comes and goes and may not occur in the course of an office visit. The physician must first rule out all possible organic sources for the weird smell, especially sinus or gum disease. The physical causes of phantosmia are diverse and include seizure, migraine, and brain tumor.

When a real odor gives rise to a distorted perception, the condition is called parosmia. The distortions in such cases are almost always unpleasant; patients say things smell foul, rotten, or burned. Such was the case of a sixty-year old woman who awoke one morning to find that every odor smelled like burnt toast. Eleven years later, despite treatment with antibiotics, antivirals, vitamins, beta-blockers, anticonvulsants, and zinc sulfate, her condition was unchanged. Most parosmics can tell you which smells are distorted; the most common are gasoline, tobacco, coffee, perfumes, fruits (mainly citrus and melon), and chocolate. Parosmia almost always occurs after an upper-respiratory-tract infection or head trauma, where smell function is reduced but not completely gone. This leads researchers to speculate that parosmia is an "incorrect rewiring" of the connections among regenerating nerve cells following damage to the olfactory system. Among smell pathologies, the most appalling is cacosmia, in which everything smells like shit.

In Philip K. Dick's sci-fi novel *The Simulacra* (1964), there is a character named Richard Kongrosian, a psionic pianist who plays the instrument telekinetically. He also has a history of mental instability. An annoying advertisement triggers in Kongrosian the delusion that he has a bad body odor. He becomes obsessed with BO and washes compulsively, but in vain; the smell lingers. His ability to play the piano from a distance notwithstanding, Kongrosian is a poster child for a real-life psychiatric disorder known as olfactory reference syndrome, which is characterized by persistent hallucinations of body malodor.

IT PROBABLY COMES as no surprise that men and women differ in smell ability. This has been confirmed many times with a variety of test methods and in cultures around the world. Women rate themselves as having a better sense of smell, and the data back them up. Women detect odors at lower concentrations and are better able to identify them by name. A German psychologist found that men and women are equally good at remembering colors and musical tones, but women are better at remembering smells. Humorist Dave Barry's wife would not be surprised:

> At least five times per week, my wife and I have the same conversation. She says: "What's that smell?" And I say, "What smell?" And she looks at me as though I am demented and says: "You can't SMELL that?" The truth is, there could be a stack of truck tires burning in the living room, and I wouldn't necessarily smell it. Whereas my wife can detect a lone spoiled grape two houses away.

Sex differences are based on group averages; there is much variability within each sex, and large overlap between them. But in general, women are better. Or, as Dave Barry put it, men suffer from Male Smelling Deficiency Syndrome.

What explains the female superiority? There is little evidence of sex differences in the nose. Dave Barry's nose probably looks and operates much like his wife's. The brain is a different story. Recent evidence suggests that brain structures related to odor perception differ in size and cellular architecture between men and women. Whether these anatomical variations explain Barry's quip remains to be seen. We do know that some male-female differences in perception (the fact that women often rate smells more intense and unpleasant) are mirrored by differences in the underlying brain-wave response.

Female smell superiority is partly due to women having higher verbal fluency; verbal skills boost performance on tests of odor memory and odor identification. Another factor is hormones. A woman's smell sensitivity varies across her menstrual cycle and is greatest around the time of ovulation. Hormone effects are not simple; they interact in complex ways with cognitive factors. This interaction produces one of the most dramatic olfactory sex differences ever observed in the lab. Sensory researchers Pam Dalton and Paul Breslin tested men and women for their sensitivity to a specific odor. With repeated testing over the course of thirty days, the women became much more sensitive to the odor, while men did not. The effect was confined to the tested smell; sensitivity to a different odor did not change for men or women. The enhanced sensitivity can't be attributed to practice; the women weren't getting better at threshold tests in general. They became more sensitive because they paid close attention to low levels of odor while being exposed to it multiple times. Most remarkably, Dalton and Breslin didn't find enhanced sensitivity in prepubescent girls and postmenopausal women. The phenomenon is limited to women of reproductive age. This implies that female hormones are needed to make it happen, and in fact it can be observed in postmenopausal women who take hormone replacement therapy.

Sex differences are evident within days of birth: baby girls turn toward novel odors and spend more time smelling them than baby boys do. The anthropologist Lionel Tiger attributes the difference to evolution. In our long history as hunter-gatherers, he says, it was the females who gathered fruits and vegetables, and a good sense of smell was valuable in judging ripeness and safety. Tiger's view—essentially a biologized version of "women spend more time cooking"—will not be received warmly in some quarters. Yet it's hard to see how a cultural explanation can explain sex differences in two-week-old infants.

WITH AGE, OLFACTORY performance begins to deteriorate. The first signs of decline are detectable in the early forties—at least under laboratory conditions—and the pace accelerates in the sixties and seventies. Interestingly, the rate of decline varies with the odor. Rose and banana, for example, are easily perceived until people are in their seventies, while mercaptans (the natural-gas warning odor) show a drop among people in their fifties. Some age-related smell loss can be traced to the nose itself—the accumulated wear and tear of infections and minor blows to the head. Some of the loss is traceable to the brain. For example, odor identification ability depends on how much short-term memory the test requires. Because short-term memory declines with age, elderly people score better when the odor test is presented in a simple yes/no format than in a multiple choice format that requires more memory capacity. In any case, decline is not inevitable; a given seventy-five-year-old may outperform a given twenty-five-year-old. Perfumers, in fact, usually get better with age. Experience and skill more than compensate for any dimming of acuity that comes with age. I know of no fragrance house with a mandatory retirement age for perfumers.

To THE AVERAGE person it seems obvious that smoking must dull the sense of smell. Surprisingly, the evidence is equivocal. Some studies find adverse effects of smoking but many, including several recent ones, do not. One, an Australian study of 942 people, found that having a smoke within fifteen minutes of smell testing put a temporary dent in performance. Other than that, "smoking did not reduce olfactory performance or self-assessment of olfactory ability in this group, contrary to previous findings." The National Geographic Smell Survey reported mixed results. For example, smokers found the artificial musk scent of Galaxolide more intense than did nonsmokers, but the reverse was true for the musky-urinous smell of androstenone. Pleasantness ratings for the skunky-smelling mercaptan sample were higher among smokers, but so were their ratings for rose and cloves. It's possible that smokers become sensitized to some odors and desensitized to others. In any case, minor effects of smoking observable in clinical testing may have little appreciable impact on everyday smell function. Indeed many perfumers, including the best in the business, have smoked like chimneys.

So strong is the conventional wisdom about the negative effect of smoking that researchers worry when they fail to confirm it. Take the case of a large population-based study in Skövde, Sweden. It linked decreased olfactory performance to several factors including being older, being male, and having nasal polyps. Smoking was not one of the factors. Similarly, diabetes and nasal polyps predicted complete anosmia, but sex and smoking did not. The authors didn't find that smoking *improved* odor perception; they merely failed to find that smoking harmed it. One can see them bracing for a wave of politically correct indignation when they say, "The lack of a statistically significant relationship between olfactory dysfunction and smoking may be controversial."

Blind Faith

When, at a party, I own up to being an expert on the sense of smell, I get peppered with questions. (I don't mind this—if I'm not in the mood for Q&A, I tell people I'm "in the chemical business" and the conversation grinds to a halt.) People often ask about smell ability. Who is better: men or women? perfumers or normal people? Curiously, one comparison doesn't come as a question but as an assertion. Wineglass in hand, someone will inform me in earnest tones that "blind people have a heightened sense of smell." Others confidently assure me that "Helen Keller had an incredibly sensitive nose."

Helen Keller has been dead since 1968, but remains a powerful symbol of the belief that blindness turns people into super-smellers by way of compensation. (The Marvel Comics hero Daredevil embodies the same idea.) Despite her iconic status, Helen Keller herself did not claim to have a supersensitive nose. In her famous essay "Smell, the Fallen Angel" she describes what she is able to smell. Amid lyrical, somewhat overripe prose ("Smell is a potent wizard that transports us across a thousand miles and all the years we have lived"), she gives specific examples of her olfactory ability. Let's compare her talents to ours. Smells trigger memories—check. Approaching rainstorms have a smell—check. Can smell if a house is old-fashioned and long-lived-in—check. Can smell a person's occupation (painter, carpenter, ironworker)—check. Close friends have distinctive odors—check. Babies smell sweet—check. Nothing extraordinary so far. Helen Keller does not sound like a nasal genius. Indeed, nowhere does she claim to have a more sensitive nose as a result of being blind, or that her sense of smell is better than that of sighted people. On the contrary, she writes, "I have not, indeed, the all-knowing scent of the hound or the wild animal." She also says, "In my experience smell is most important." It is not surprising that, being blind and deaf, she finds smell to be her primary way of sensing the world.

Helen Keller's modest assessment of her own ability has done little to dampen enthusiasm for the idea of smell compensation in the blind. It seems so reasonable it must surely be correct. But is it? There is plenty of experimental evidence that addresses the question—in the last twenty years, six studies have compared smell in the blind and the sighted. Without exception, they find that the blind are no more sensitive than the sighted—both groups detect odors at about the same concentration. Nor do blind and sighted people differ in the ability to discriminate one odor from another. Even the brain waves triggered by odor stimulation are similar in blind and sighted people.

Blind people may have one advantage: in three of the six studies, they were better at naming odors. Even here, their success depended on cognitive factors such as memory rather than hyperacute perception. Based on her own words, and on what has been observed in experiments, Helen Keller's ability to navigate the smellscape was not the result of a supersensitive nose. Rather, it was a triumph of the adaptable human brain making the most out of a perfectly ordinary nose.

The Nose on Freud's Face

Sigmund Freud was not a big fan of the nose. He believed odor perception was vestigial, the sensory equivalent of the appendix. In his view, smell became obsolete when our evolutionary ancestors took on upright, bipedal posture and put distance between the nose and the ground. At the same time, Freud's ape-man discovered shame and disgust in the exposure of his genitals. This led him to turn away from the stink of excrement and to suppress his sense of smell in general. To Freud, this was a vital precondition for the emergence of civilization—the repression of smell meant the repression of wild sexual impulses and their redirection to more refined behavior. Freud thought that children recapitulated the history of the species as they grew up, and thus that the infant's early interest in smell fell away like embryonic gill slits. Freud's leading American disciple, A. A. Brill, summarized the master's view: "All children make good use of the sense of smell in early life; some of them, as we shall learn later, retain it even in adult life; most of them, however, lose it, so to speak, as they grow older." To the orthodox analyst, psychologically mature adults move on and leave fascination with smells to perverts and neurotics.

Like many of Freud's theories, his views on smell are difficult to summarize without making them sound simpleminded and ridiculous. The original texts consist entirely of a few sentences in a letter to his confidant Wilhelm Fliess, a German ear-nose-and-throat physician, and two footnotes in the book *Civilization and Its Discontents*, and are part of what historian Peter Gay called Freud's "audacious, highly speculative venture into psychoanalytic prehistory."

Nevertheless, after becoming part of the bedrock of psychoanalytic theory, they helped devalue smell in the wider intellectual world.

It is puzzling that Freud, who found a sexual angle in every other facet of psychology, thought it had so little to do with the sense of smell. Is sexual attraction no longer an affair of the nose? Are modern women scentless and modern men oblivious, or vice versa? In a recent University of Texas study, men said T-shirts worn by women near the time of ovulation smelled more pleasant and sexy than T-shirts worn during a nonovulatory part of the cycle. Modern women, it seems, continue to produce a scent cue associated with ovulation and modern men continue to respond to it. This low-technology experiment could have been done in Vienna in 1930 or New York in 1932, had either Freud or Brill cared to test their theories.

Brill toed the party line when he wrote in 1932 that "the sense of smell unlike the sense of sight plays a very small part in the life of civilized man," and "modern man has little need of his sense of smell." Though surrounded by modern, civilized men, Freud and Brill never bothered to ask their opinion. The psychologist Paul Rozin and colleagues got around to it a few years ago. They asked people to rank the unacceptability of permanent loss of sense of smell, loss of hearing in one ear, and loss of the left small toe. According to about half the respondents, loss of smell was the least acceptable alternative. The average person is not as dismissive about the sense of smell as Freud thought he was. What could have motivated Freud to construct a psychoanalytic conjecture so flimsy it could be blown up with a simple opinion poll?

The experts think it was something, well, Freudian. The psychoanalyst Annick Le Guérer attributes it to Freud's "repression" of his "transferential relationship with Fliess." The anthropologist David Howes thinks Freud's conflicted emotions toward Fliess led to his "denial of nasality" and a desire to "cut the nose out of psychoanalytic theory."

I have a more straightforward hypothesis. Based on the facts of his medical history, I suspect Freud suffered from hyposmia. The repeated insults of cocaine, nose surgery, influenza, sinus infection, cigar smoking, and finally aging left him with a clinically impaired sense of smell.

Freud caught influenza in the spring of 1889, at the age of thirty-three. The infection was severe enough to leave him with a persistent cardiac arrhythmia, so it could easily have affected his nose. In his letters to Fliess from 1893 to 1900, Freud often complains of nasal congestion with discharge of pus and scabs, both symptoms of sinus and nasal passage infection. Freud suffered from migraine headaches, which he treated with nasal applications of cocaine prescribed to him by Fliess. Fliess operated on Freud's nose twice to remove and cauterize part of the turbinate bones. On top of all this, Freud smoked heavily; his typical rate in the 1890s was twenty cigars a day.

Freud's nose was already a medical disaster zone when he hatched his smell theory in 1897, and my hunch is that he was already smell-impaired. When he wrote *Civilization and Its Discontents* in 1930, he was seventy-four years old and suffering from cancer of the jaw as well. In my view, Freud's intellectual indifference to smells was the result of sensory deprivation—the gradual onset in adulthood of severe hyposmia. His ludicrous idea that smell was active in children but ceased to matter for adults had nothing to do with his feelings about Wilhelm Fliess. It was simply an overgeneralization of his unfortunate personal experience.

Our Rank in the Animal Kingdom

No doubt there is a vast difference in power in the sense of smell in both these animals [deer and dog] and in man; nevertheless, I do not think so meanly of man's olfactories as some physiologists appear to do.

—W. H. Hudson, *On the Sense of Smell* (1922)

After finishing my PhD at the University of Pennsylvania, I began working a few blocks away, at the Monell Chemical Senses Center. I received a fellowship to study there with Dr. Kunio Yamazaki. He had several lines of inbred mice used for cancer research; the lines were genetically identical except for a set of genes known as the Major Histocompatibility Complex (or MHC), which controls the body's tissue-rejection response. They are the genes used to find whether a person is a suitable match as an organ donor. Yamazaki's mice preferred to mate with individuals bearing a different MHC type apparently on the basis of smell. My plan was to study the behavior behind the odor-based mate choice using competitive mating experiments, where the female had access to multiple males of different MHC types.

Watching the mice choose mates, I became curious. Could humans detect the odor differences that were so apparent to the mice? Soon I was running my first experiment on human odor perception. I had blindfolded people sniff live mice in Tupperware containers with holes cut in the sides. Occasionally a mouse tail would get up someone's nose; this seemed to bother some people more than others. The judges also sniffed tiny test tubes filled with mouse urine or dried fecal pellets. (Thankfully, no one inhaled a mouse turd.) For every odor source the results were clear: untrained humans could distinguish between the mouse strains based on smell alone. The uncanny scent powers of mice were well within human reach. I wrote up the results as a man-bites-dog story for the *Journal of Comparative Psychology*, and it eventually became one of the most-cited scientific papers I have ever published. By encouraging me to continue exploring human odor perception, it also led to my career in the perfume industry.

Deborah Wells and Peter Hepper discovered an even more impressive man-smells-dog story. They had dog owners sniff two identical blankets, of which one had been slept on by their pet and the other by an unfamiliar dog. The owners correctly identified their dog 89 percent of the time. The strength or pleasantness of the smell was not a factor, nor were non-doggy household odors.

Stories about the amazing ability of the canine nose highlight the dog's talent and ignore how the feat is stage-managed by humans. ("Pay no attention to that man behind the curtain!") Consider the recent finding that dogs can sniff out bladder cancer. The dogs in question were trained exhaustively with human urine samples. Training began with search-and-find games and progressed to more complex tests. Urine samples were carefully selected so the dogs would learn to ignore irrelevant dietary odors. The trainers also counterbalanced samples from smokers and nonsmokers, patients and healthy people. After seven months of training, the dogs were ready for the decisive test: to pick the single positive sample from a set of seven. As a group they were correct 41 percent of the time, which successfully beat the random odds (which were one in seven, or 14 percent). The resulting scientific report made headlines around the world.

So, yes, dogs can smell odors associated with bladder cancer. But this is a far cry from "What's that, Lassie? Timmy has bladder cancer?" To make use of this canine talent, your local hospital would have to maintain a half-dozen dogs and their trainers, supply copious medically certified human urine samples, and provide ongoing statistical support and chemical analysis. At which point six out of ten bladder cancers would go undetected.

If the human nose received the same gee-whiz treatment given to animal stories, we would sound as impressive as any dog. Here's an example: Just by smelling some ice cream that once had a wooden popsicle stick in it, regular folks can tell whether the stick came from Wisconsin, Maine, British Columbia, or China. Amazing, no? How do those monkey-people do it? In this case, wooden sticks from each locality were frozen in vanilla ice cream for six days. The samples were melted and the sticks removed. The sniffing primates—Ohio State graduate students—had to pick the same sample from a repeatedly presented pair of samples five times in a row to be declared a success. All possible pairs of wood source were tested. Two judges failed—they couldn't tell one stick-scented ice cream from another. Eight judges succeeded—they could reliably discriminate anywhere from five to nine of the ten possible pairings. Not bad for humans. Could the judges explain how they did it? Unfortunately not, but then, neither could the cancer-sniffing dogs.

The physicist Richard Feynman had a great party trick in which he would identify by smell objects briefly handled by other guests when he wasn't looking. He said it was easy to do because peoples' hands have surprisingly different scents. (A 1977 study confirmed that hand odor is individually distinctive and discriminable.) There are other stupid human tricks besides Feynman's. For example, in a lineup of dirty laundry we can pick out the T-shirt worn by our spouse or partner. A mother can identify the smell of her own baby, and a baby can pick out the scent of its mother's breast.

How do humans measure up at the quintessential doggy task—scent tracking? Researchers at UC Berkeley had people get on their hands and knees and follow a 10-meter-long chocolate-scented trail using only their noses. The test subjects wore goggles, gloves, and kneepads, which restricted nonolfactory input. Two-thirds of the people tested successfully followed the trail under these conditions. (When subjects wore nose plugs, none of them could follow the trail.) With a few days of training, tracking speed was doubled and people strayed less from the scent trail. Dog lovers (of which I am one) may also be surprised to learn that drug dogs and humans have almost identical sensitivity to methyl benzoate, the smell used to track cocaine. Dogs have great noses, but it's time to stop the trash talk and give ourselves more credit.

Many people take it for granted that the human nose is inferior, and scientists often make the same assumption. Charles Darwin thought our evolutionary ancestors made good use of smell, but felt it was "of extremely slight service, if any" to modern man. The sex psychologist Havelock Ellis agreed: "Among the apes it has greatly lost importance and in man it has become almost rudimentary, giving place to the supremacy of vision." The attitude persists. As recently as 2000, some French researchers asserted "The sense of smell in primates is greatly reduced (microsmatic) with respect to other mammals such as dogs or rodents."

Scientists are taking a fresh look at the conventional wisdom regarding the sense of smell in animals. The anatomists Timothy Smith and Kunwar Bhatnagar, for example, are questioning the textbook distinction between macrosmatic and microsmatic animals, i.e., those with good and poor olfactory abilities. The long-standing assumption is that what separates macrosmatic and microsmatic species is the amount of surface area inside the nose. This turns out to be a bad assumption; internal surface area is more about air conditioning—warming and filtering incoming air—than about odor perception. Of more relevance is the amount of sensory tissue in the nose. But Smith and Bhatnagar find that the amount of sensory tissue varies from species to species independently of total surface area. Further muddying the waters, the number of olfactory nerve cells per square inch varies among species. All in all, surface area is a dubious proxy for smell ability. Smith and Bhatnagar suggest that the traditional macrosmatic/microsmatic distinction has outlived its usefulness. Size isn't everything.

The Yale University neurobiologist Gordon Shepherd agrees that counting nerve cells is a poor way to estimate sensory talent. In his view, the number of cells available for odor detection is less important that what the brain does with the information those cells provide. He makes the analogy to hearing: humans have about the same number of auditory nerve fibers as cats and rats, yet we have far superior speech abilities. It's the brain areas that analyze and interpret speech sounds that provide the advantage, not the number of cells in the ear.

The German sensory physiologist Mathias Laska cuts right to

the chase by measuring odor perception in different animal species. He has used reward-based conditioning techniques to find odor detection thresholds in spider monkeys, squirrel monkeys, and pig-tail macaques. According to conventional wisdom, these primates are less sensitive than dogs and rabbits, but Laska finds they perform quite well—monkey thresholds are comparable to those of dogs and rabbits across a variety of odors. And contrary to Darwin's gloomy belief, Laska finds that humans have odor sensitivity similar to that of apes and monkeys.

New evidence suggests that humans and animals may be more similar in odor perception than we thought. In 1991, Linda Buck and Richard Axel discovered a large family of mammalian olfactory receptor genes, work for which they eventually received the Nobel Prize. Each gene produces a different receptor. In general, more receptors means more detectable odors, and therefore greater smell ability. Rats have about 1,500 functional receptors, followed by dogs with about 1,000, mice with about 900, and chimpanzees with about 350. Humans have somewhere between 340 and 380. Dolphins have zero.

Does this mean rats are five times better smellers than we are? Not really. We can use DNA sequence similarity to arrange odor receptors into families and subfamilies. In theory, similar receptors detect similar odor molecules, so a receptor subfamily detects a class of related odors. When we compare odor receptor subfamilies, the human-animal gap doesn't look too large. Humans and dogs have about 300 subfamilies, rats have 282, and the mouse 241. The over-lap between species is substantial. About 87 percent of human receptor subfamilies have counterparts in the mouse genome, while 65 percent of mouse subfamilies are shared by humans. This sug-gests to Linda Buck and her colleagues that "the majority of odorant features [i.e., smells] detectable by one species may also be recog-nized by the other." Perhaps a mouse can smell more of our world than we can smell of his. (Unlike us, he may have a whole subfamily of receptors devoted to cat urine.) For man and mouse the differences are not as big as the similarities. For man and chimp this is even more the case—there is a corresponding human gene for 85 percent of chimp odor receptor genes. Chimp, dog, man, or mouse, we perceive the general features of the smellscape in much the same way.

Physical equipment—size of brain areas, number of nerve cells or receptor types—may be less important than what the brain does with the information once it arrives. For many animals, a smell is a call to action, a trigger for a biologically hard-wired survival response: "scent of lion means flee." In contrast, human cognitive abilities turn smells into symbols and let us make flexible use of their signal value. When it comes to comparative smell ability, it's the brain, stupid.

Better Than the Rest

> One morning when I walked on my monk's alms-rounds to collect food, my nose became like that of the most sensitive dog. As I walked down the street of a small village, every two feet there was a different smell: something being washed, fertilizer in the garden, new paint on a building, the lighting of a charcoal fire in a Chinese store, the cooking in the next window. It was an extraordinary experience of moving through the world attuned to all the possibilities of smell.
>
> —JACK KORNFIELD, *A Path with Heart*

My friend Larry Clark is an ornithologist. I have hiked trails with him as he identified bird after bird by song alone. His skill leaves me awestruck. It's the same feeling I get when a perfumer talks about a fragrance—he seems to be smelling more than I do, finding notes that my blundering nose doesn't detect until he points them out.

How do olfactory experts accomplish these feats? Are their noses that much better than yours or mine? What does it take to become an expert smeller?

Pure nose-sensitivity is not the answer. The average person probably detects odors at about the same concentration as the professional wine taster. What the expert has are cognitive skills that make better use of the same sensory information. The practiced wine expert can name varietals and tell one vintage from another, just as the trained perfumer classifies a new cologne with ease and zeroes in on its unique notes. The expert's advantage, in other words, is brain power rather than nose power, and in the regular exercise of these specialized mental skills. Wine experts, for example, routinely make notes as they taste. Experts outperform novices in matching their own descriptions to wines on subsequent tastings. Their mental discipline helps experts avoid a trap called the "verbal overshadowing effect" that can snare novices when the effort to generate a verbal label interferes with the perception of the aroma itself.

The perfumers Robert Calkin and Stephan Jellinek believe their job can be done with only an adequate nose. What makes for professional success is specific mental skills and thought processes. My own research confirms that fragrance professionals think differently. Perfumers, fragrance evaluators, chemists, and sales executives have better olfactory imagery ability than nonexperts from outside the industry. The ability to bring to mind the scent of specific perfumes, and to imagine how ingredients will smell when blended, is central to the job description.

Constant honing of perceptual skills may actually change how an expert's brain responds to scent. The brain-wave patterns of professional perfume researchers have been compared to those of less specialized workers. When smelling an odor, the pros show distinctive frontal lobe activity in an area known as the orbitofrontal cortex— one that is involved in cognitive judgments. This pattern of brain response in the pros may reflect their more analytic way of perceiving odor. Another study examined brain activity in wine sommeliers and nonexperts, as each group sipped and savored wine samples. The sommeliers had activity in areas associated with cognitive processing (the orbitofrontal cortex again) and in an area where taste and smell information are integrated. In contrast, the nonexperts showed activity in the primary sensory areas and zones associated with emotional response. Practice in making deliberate judgments about what one smells leads to changes in brain function and makes a person into a better smeller.

Superpowers

Is there such a thing as an olfactory prodigy? What talents would a Mozart of the nose possess? He would ace tests of odor identification, notice smells at trace concentrations, and quickly zero in on the difference between highly similar scents. He would effortlessly arrange samples according to concentration, name odors without hesitation, and pick the individual components out of a complex mixture. He would have an enormous store of remembered odors, and the ability to memorize new ones in a single sniff. He couldn't be fooled into false recognition with distracters and decoys. Finally, he would have a profound ability to imagine odors and to anticipate how they would smell when mixed together.

If such a person exists, science hasn't found him. This doesn't stop novelists from imagining characters endowed with superhuman ability. Take Grenouille, the hero of Patrick Süskind's novel *Perfume: The Story of a Murderer.* Many people have recommended this book to me, thinking I would enjoy the depiction of Grenouille's incredible olfactory powers, but I am not impressed. Where the Laing Limit keeps normal people from smelling more than four odors in a complex mixture, Grenouille is born with the ability to recognize dozens of them. Even if we buy this fantasy, how does it instantly make Grenouille the best perfumer in Paris? Analyzing a perfume isn't the same thing as creating one. (I can hear every note in a Mozart symphony, but that doesn't make me a composer.) We know that perfumers work from the top down; they first recognize a perfume's type, and then the nuances that make it unique. Grenouille starts by cracking a perfume into its raw materials, the very opposite of how real perfumers work. As a fan of slasher films, I don't mind that Grenouille is a repellent freak with no body odor of his own who murders female virgins to extract their body scent. Neither do fans of *Perfume*—they are so enthralled by the romance of essential oils and blending that they ignore Grenouille's nasal necrophilia and the novel's soul-deadening grimness. *Perfume* is about perfume-making the way *The Texas Chainsaw Massacre* is about sausage-making.

The novelist Salman Rushdie created a hero with supernormal olfactory power named Saleem Sinai, who is born with an enormous nose. The smelly passages in *Midnight's Children* are fun to read even as they verge on the phantasmagorical. Here Rushdie conjures the smellscapes of Karachi, Pakistan:

> . . . the fragrances poured into me: the mournful decaying fumes of animal faeces in the gardens of the Frere Road museum, the pustular body odours of young men in loose pajamas holding hands in Sadar evenings, the knife-sharpness of expectorated betel and opium: "rocket paans" were sniffed out in the hawker-crowded alleys between Elphinstone Street and Victoria Road. Camel-smells, car-smells, the gnat-like irritation of motor-rickshaw fumes, the aroma of contraband cigarettes and "black-money," the competitive effluvia of the city's bus-drivers and the simple sweat of their sardine-crowded passengers.

Like Grenouille, Saleem Sinai comes from the land of make-believe. His olfactory ability goes far beyond normal experience: he uses it to detect emotions in other people, read their character, and peer directly into their souls. Similarly bizarre characters appear in Chitra Banerjee Divakaruni's *The Mistress of Spices* and Tom Robbins's goofy burlesque *Jitterbug Perfume*. Why are the authors of magic realist fiction so fond of supersmellers? Transforming a "primitive" animal sense into an all-knowing form of perception is apparently an irresistible literary conceit. Whatever their entertainment value, fictional supersmellers don't shed much light on real people.

Busted

> I suggest that if the police really wish to know where stills and "speakeasies" are located, they take me with them. It would not be a bad idea for the United States Government to establish a bureau of aromatic specialists.
>
> —HELEN KELLER

In April 2005, an Indiana man arrived at the Decatur County jail to bail out his brother-in-law. As he handed over $400 in cash, the dispatch clerk noticed that the bills reeked of marijuana. Police officers got the man's permission to search his car. They found a pipe and some pot and charged him with possession. The episode has a certain Cheech-and-Chong quality to it, but the use of odor as evidence raises serious questions about the Fourth Amendment's guarantee against unreasonable search and seizure. In February 1999, an Ohio State Highway Patrol officer stopped a motorist for running a red light. When the driver rolled down his window, the officer smelled marijuana smoke. A search turned up rolling papers and joints in the driver's pocket, and a stubbed-out doobie in the ashtray. The driver was arrested. At trial, he succeeded in having the charges dismissed on the grounds that a search based only on odor—without other visible, tangible evidence—was improper. The case made its way to the Ohio Supreme Court, which ruled that the "plain smell" of burning pot was, by itself, sufficient probable cause for a warrantless search. The supreme courts of Michigan, Colorado, Wisconsin, Arkansas, and at least fifteen other states have reached similar conclusions.

How good a nose does a government agent need to claim probable cause for a drug search? The Ohio court relied on the fact that the arresting officer was trained and experienced in identifying the smell of marijuana smoke. Other jurisdictions aren't so fussy. The degree of nasal prowess claimed by police officers can, at times, beggar belief. In New Jersey, for example, a driver was pulled over for a traffic infraction. The police officer claimed to smell fresh, unburned marijuana through the open driver's window. A search revealed a brick of Mexican pot wrapped in a plastic garbage bag in the trunk, where it had been placed after a drug buy twenty minutes earlier. In California, police searched a house where they suspected pot was being grown. They didn't obtain a warrant because they claimed they could smell marijuana plants—from several hundred yards away in the hot air and diesel exhaust venting from the chimney.

These feats of nasal detection are all the more remarkable given the level of training most police officers receive. According to Jim Woodford, who serves as an expert witness in criminal trials, officers often learn drug smells by sniffing the real thing in the evidence room. Formal training is rudimentary. "Somebody comes in with a suitcase of stuff, everybody goes by and takes a sniff. That's the training," he says. The problem with this informal approach is that marijuana aroma is extremely variable, something potheads are well aware of. (Just ask the reviewer of the Beck concert in Costa Mesa . . .)

Of course, police officers become familiar with drug smells while busting dealers and users. They cite this on-the-job experience when defending their skills in the courtroom. They testify that "I've been on so many busts, and I recognize it. Over the years I've learned it."

Woodford says, "That's sufficient to be deemed an expert by the court." He says it is rare for the defendant in a drug case to challenge the officer's smell ability via a smell test or medical exam.

Just how detectable is the smell of pot under circumstances such as these? Richard Doty and colleagues conducted some forensic sniff tests to find out, using experimental conditions modeled on the New Jersey and California cases. They found that untrained people can easily distinguish a Hefty bag containing 2.5 kilograms of Mexican pot from one holding an equal weight of shredded newspaper. But when the samples were placed in a car trunk, the panelists could not detect the smell from the driver's window. Likewise, panelists could reliably identify mature female cannabis plants at close range by scent alone, and could distinguish immature pot plants from tomato plants by smell. But when the smell of marijuana plants was mixed with exhaust from a diesel generator, the panelists couldn't detect it.

When it comes to detecting drunk drivers, sniff-based forensics are on even shakier scientific ground. A study by the National Highway Transportation Safety Administration found large variability in the ability of police officers to smell alcohol on a person's breath. As a group, cops picked up the scent consistently only when the drinker had a very high blood alcohol level (the detection rate was 61 percent for BACs between 0.10 and 0.15 percent). In the most rigorous study on the topic, all variables except odor were eliminated: test subjects were hidden behind a screen and breathed at the officers through a tube. The police participants were all highly experienced and trained as Drug Recognition Experts. Even so, test performance was highly variable across officers. As a group, they detected breath alcohol 85 percent of the time at BACs of 0.08 percent or more, but caught it only two-thirds of the time at lower levels. An officer's ability to estimate the intensity of breath alcohol odor was no better than chance.

THAT POLICE OFFICERS, like everyone else, show a wide range of olfactory ability comes as no surprise to smell scientists. That their abilities should be granted special consideration by judges and juries is another matter. Doty and his colleagues argue that skepticism is in order when marijuana is said to be "in plain smell." Sensory claims by police are least substantiated when it comes to fresh, unburned marijuana. Yet this is just the circumstance where courts have given greatest credence to a police officer's nose—no corroborating evidence is needed. Doty's study has already been cited by the defense in a drug case in federal court. (A police officer with no training in pot aroma claimed to smell immature marijuana plants in an unvented grow house from a long distance away.) Can trained police officers outsniff civilians? Probably. But according to Doty, this has yet to be scientifically documented. Helen Keller would expect better from the Federal Bureau of Aromatic Specialists.

CHAPTER 4

The Art of the Sniff

The smoke of my own breath;
Echoes, ripples, buzz'd whispers, love-root, silk-thread,
   crotch and vine;
My respiration and inspiration, the beating of my heart,
   the passing of blood and air through my lungs;
The sniff of green leaves and dry leaves, and of the shore, and
   dark-color'd sea-rocks, and of hay in the barn . . .
   —WALT WHITMAN, *Leaves of Grass*

SOME SMELLS ARE MORE SUBTLE THAN OTHERS. THEY float up the nose on the tidal rhythms of normal breathing and may not reach conscious awareness until minutes later. When we want to pay attention to an odor, we don't wait for the next lungful of air—we capture it with a sniff. Sniffing is an odd behavior—it has no analog in vision or hearing. (Dogs, mice, and deer can rotate their external ears to focus on sounds; we can't.) Sniffing is ignored by students of "body language." It can be done covertly, and in polite company it usually is; sniffing is considered rude, and audible sniffing is downright vulgar. It takes an uninhibited, bumptious soul like Walt Whitman to draw attention to it, much less revel in it. But there is no getting around it; sniffing is essential. Whether one is tracking down a dead mouse in the basement or savoring a newly opened bag of Doritos, the sniff is the prelude to a smell.

The purpose of a sniff is to get scent molecules to the place where we can smell them. The question that took philosophers and scientists thousands of years to answer was, Where exactly does smelling happen? Some ancient Greek philosophers argued that it took place in the nose, but the sievelike appearance of the cribriform plate—a bone at the base of the skull just above the nasal passages—led others to speculate that odor particles made their way directly to the brain through these tiny holes. In this view, the nose is a merely a tube and the brain is the sensory organ. The ancient nose-versus-brain debate wasn't settled until 1862, when a German anatomist discovered the olfactory nerve cells in a cleft high in the nasal passage. Smell—at least the first physiological contact with odor molecules—clearly happens in the nose. The holes in the cribriform plate are there to allow nerve fibers from the sensory cells to reach the brain.

Because the olfactory cells were tucked away in a narrow olfactory cleft, they did not appear to be exposed to the main flow of air through the nose. Researchers were soon asking how much of air entering the nostrils actually made it to the olfactory nerve endings. Early experiments were ingenious and also a bit macabre. In one study, for example, the head of a cadaver was cut in half and tiny squares of litmus paper were placed throughout the nasal passages. The head was reassembled and ammonia vapor pumped through the nostrils and out the trachea. Color changes in the papers showed that very little ammonia-laden air made it to the sensory cells; most passed through the lower passages. A second, more grotesque experiment anticipated the slice-and-shock art of Damien Hirst by a century. A split cadaver head was pressed against a glass plate and smoke was blown into the nostril. Observers could see the currents and eddies as the smoky air flowed through the complex folds of the nasal chamber. The smoke patterns, like the ammonia vapor, showed that only a fraction of the incoming air made it to the receptors.

Today, sophisticated computer models can simulate nasal airflow. Researchers can see where the flow is laminar (smooth) and where it is turbulent. They can calculate how many scent molecules are deposited onto the sensory surface as air is drawn across it. For all the high-tech apparatus and numerical precision, the modelers reach the same conclusion as their head-splitting predecessors: only about 10 percent of inhaled air blows across the nerve endings in the olfactory cleft.

THE SNIFF—a short inhalation with a high rate of airflow—is an essential step in odor detection. By forcing more air past the olfactory cleft, we take a bigger sample of the external smellscape. So how did it come to be dismissed and even suppressed by serious scientists? This is a strange tale. The first scientist to pay much attention to sniffing was also the one who tried to eliminate it from smell experiments. In 1935, Charles A. Elsberg was a highly regarded neurological surgeon in New York with a flair for invention— he designed surgical instruments and had performed the first successful removal of a herniated spinal disk. Elsberg's flair for promotion was even bigger. He had cofounded the Neurological Institute of New York, set up the country's first Neurosurgery Service there, and later cofounded the Society of Neurological Surgeons. At the age of sixty-four, it occurred to Elsberg that brain tumors, by exerting pressure on the olfactory areas at the base of the brain, might lead to impaired odor perception. If he could measure odor sensitivity, he might be able to identify patients with brain tumors. Accordingly, he came up with a method that involved a bottle, a cork, a syringe, and some rubber tubing. The patient would hold his breath and Elsberg would inject odorized air into his nostril. Acuity was measured by how big a blast of air was needed for the patient to detect a smell. Elsberg found that a normal person needed six to nine cubic centimeters' worth. Elsberg's system was coldly efficient; it not only eliminated sniffing, it eliminated breathing.

Elsberg touted his method as a major breakthrough: the first scientifically objective measurement of odor sensitivity. He either didn't know of, or didn't care to acknowledge, the olfactometer invented thirty years earlier by Hendrik Zwaardemaker. Every sensory psychologist in America was familiar with Zwaardemaker's device, and most had one in the laboratory. It consisted of a glass sampling tube, curved at one end to fit into a nostril. A wider tube, containing an inner layer of scented material, fit snugly over the sampling tube. The farther the wide tube was pulled back, trombone-like, off the end of the sampling tube, the more scented surface was exposed. Sensitivity was measured as the length, in centimeters, that the scent tube had to be withdrawn in order to create a detectable level of odor. Zwaardemaker's device, of which several versions were available, was reliable enough to explore the basic phenomena of odor perception and was used in laboratory demonstrations in colleges across the country. Nevertheless, Elsberg's results were soon written up in *Time* magazine and on the front page of the *New York Times*. In the latter, the headline read, "Brain Tumors Detected by Scent with Device Keener Than the X-ray; Neurologists Hail Dr. C. A. Elsberg's Discovery as Epochal—Based on Accurate Measurement of Sense of Smell, Which Was Viewed as Impossible Heretofore." According to the credulous report in the *Times*, "Dr. Elsberg succeeded for the first time in measuring what had hitherto been considered universally as unmeasurable. He established a definite 'scent yardstick.'"

Having nine cubic centimeters of air rammed up one's nose is no barrel of laughs. However, blast injection proved to be a popular technique: most scientists prefer tight experimental control, even when precision comes at the cost of realism. Eventually researchers grew skeptical about the Elsberg method. They found that blast volume mattered less than blast force—this undercut the use of volume as a measure of smell ability. Even more troublesome, blast force was irregular—it depended on how abruptly the experimenter released the pinchcock on the rubber tube. The enthusiasm for nostril-blasting ended in 1953 when a psychology professor at UCLA compared odor sensitivity measured by Elsberg's method and by natural sniffing. Blasting produced unreliable data, while natural sniffing produced very reliable data. The results blew Elsberg out of the water. Blast injection was not the scent yardstick he claimed it was. As the syringes and hoses were packed away for good, another psychologist ruefully wondered whether "we might be better off today if Elsberg had never publicized his creation."

Mr. Natural: Keep on Sniffing

The physical characteristics of a sniff are smell dependent. Confronted with a weak scent, we take larger and longer sniffs, and more of them. We take smaller, shorter, and fewer sniffs to a strong odor. Considering how essential sniffing is to smelling, one might think this behavior would be studied by many scientists. Yet the bulk of what we know about sniffing is largely thanks to the work of one person, the Australian psychologist David Laing. He pioneered the natural history of the sniff.

In a series of elaborate studies beginning in 1982, Laing established how the dynamics of sniffing relate to smell. He controlled what people smelled with an air-dilution olfactometer, a device that generated a stream of air with precisely controlled odor levels. He measured how they sniffed by means of an oxygen mask with a tiny airflow probe concealed in it.

Laing found that natural sniffing took place in an episode of three and a half sniffs on average; some people used fewer, some many more. A person's sniff episodes have a characteristic pattern that is stable across different odors and tasks. Sniff patterns were so stable and individually distinctive that Laing found he could identify a person by airflow data alone. He went so far as to liken sniff patterns to fingerprints.

At the time of Laing's work, I was beginning my first experiments on human odor perception at the Monell Chemical Senses Center in Philadelphia. My odor sources were plastic squeeze bottles with fliptop caps. I would sit behind a screen and hand one bottle at a time to my test subject, who would squeeze, sniff, and rate the odor. As I listened to the wheezing of the bottles, I realized each person had a typical sniffing style. I soon developed a private taxonomy of sniffers. There were the Delicates, who took tiny, barely audible sniffs. There were the Honkers—people who squeezed the hell out of the bottle and inhaled so forcefully I thought they might hurt themselves. I also observed different psychological profiles. There were Decisives —people who sniffed and promptly announced their rating—and there were the Agonizers, who sniffed and resniffed and sniffed again before summoning up a rating. Every combination of behavior and decision-making style turned up in my lab: Delicate sniffers who were very decisive, Honkers who were Agonizers, and so on. These patterns were so consistent that after two or three squeeze bottles I could predict how long the entire test would take. A diverse range of local oddballs answered our recruiting ads. Once, in the middle of a test, my research assistant handed a sample of patchouli around the screen. There was some squeezing and sniffing, followed by a long silence. Finally she looked around to find that her subject had poured the sample into his hand and was massaging it into his beard. He said he liked how it smelled.

Intuitively, it seems the more one sniffs, the better one smells. Like dogs at a fire hydrant, multisniffers must be extracting every last bit of information from a smell. But are they? David Laing systematically controlled sniffing to see how it affected a person's ability to detect and describe a smell. Sometimes he allowed his subjects to sniff with their natural pattern; other times he told them exactly how many sniffs to take, how long to wait between sniffs, or how big a sniff to take. When subjects were limited to a single sniff, they took one that resembled the first in a natural sniffing episode. Whether the sniff was the first-and-only or the first-of-many, it did not appear to vary with odor strength. After many experiments he could state his findings in a nutshell: "a single natural sniff provides as much information about the presence and intensity of an odour as do seven or more sniffs." A natural first sniff can't be beat. (For the technically minded, the optimum sniff has an inhalation rate of 30 liters per minute, a volume of 200 cubic centimeters, and a minimum duration of .40 to .45 seconds.)

There are two aspects to sniffing that are reflected in how we use the verb "sniff." It can refer to a purely mechanical act (the drawing of air "through the nose with short or sharp audible inhalations") or to an olfactory experience ("to smell with a sniff or sniffs"). The dictionary's dichotomy between physical and sensory sniffing is programmed into the central nervous system at a profound level. The brain is not a passive recipient of smells drawn up the nose; it actively manages the acquisition of odor by the nose, and it does so on a time scale of milliseconds.

UC Berkeley smell researcher Noam Sobel was puzzled to find smell-related activity in the cerebellum, a brain area principally involved in tactile discrimination and the control of motor movements. When he and his lab team followed up, they discovered that two parts of the cerebellum were involved in sniffing. One was a smell-activated area; it lit up when a person smelled an odor. The stronger the odor, the greater the activation. Normally this area is activated in the course of sniffing scented air. Sobel found it was also activated by passive smelling, where odors were puffed into the subject's nose through a tube while they held their breath. The second area of the cerebellum is sniff-activated; it lights up during the physical act of sniffing, but not during passive smelling. The sensation of air flowing through the nose explains the activation in the tactile part of the brain. When topical anesthetic was applied to a subject's nasal passages to numb the nose, brain activity plunged. Together, two brain areas adjust sniff size to odor strength. This feedback happens very quickly: less than two-tenths of a second into the sniff. (By measuring with far greater precision than was available to Donald Laing, Sobel's group found that the first sniff of a series was not fixed—only its first 160 milliseconds were.) As a strong odor is detected, the cerebellum signals the respiratory muscles to throttle back on the sniff. What appeared at first to be anomalous brain activity led Sobel and his team to a new understanding of how the brain shapes our perception of smell. The cerebellum is doing what it excels at: monitoring sensory input (in this case odor strength), in order to control a motor action (inhalation).

So CLOSELY IS sniffing tied to odor perception that people routinely sniff when they are asked to imagine a smell. Without prompting, they take larger sniffs when imagining pleasant odors and smaller ones when imagining malodors. During visual imagery the eyes explore an imagined scene using the same scan paths made when viewing the actual visual scene. Preventing eye movements during visual imagery—by having people stare at a stationary target— reduces the quality of the image. Sobel found that, similarly, imagined odors were more vivid when people could sniff than when they were wearing nose clips and unable to sniff. Actually sniffing increased the unpleasantness of an imagined bad smell (urine) and increased the pleasantness of a good one (flowers). Sniffing at an imaginary odor isn't an absentminded habit—it's a behavior that improves the mental image we are trying to create. Sobel's claim that "the sniff is part of the percept" would have outraged Charles Elsberg, but it sounds reasonable to most neuroscientists today.

We have in fact done a complete about-face since Elsberg's attempt to measure smell without sniffing. Because smelling *is* sniffing, we can now test odor perception by measuring sniffing alone. We can take advantage of the fact that people naturally and unconsciously take smaller sniffs when an odor is present: the stronger the odor, the smaller the sniff. People with no sense of smell fail to adjust; they keep inhaling as if the air were unscented. A new smell test, developed by University of Cincinnati psychologists Bob Frank and Bob Gesteland, is simplicity itself. The patient wears a pair of standard-issue medical nose tubes connected to an electronic console, and sniffs at half a dozen cylinders in a row. That's it—test over. No need to identify smells by name, no multiple-choice questions, no rating scales, no fancy odor generators. Here's how it works: Each cylinder is the size of a can of beans and may or may not contain a slightly unpleasant odor (in pilot testing, Frank and Gesteland used methylthiobutryate, which has the character of feces, putridity, decay). The test console records airflow into the patient's nose and computes the size of each sniff. It compares sniffs made when the patient was smelling scented cylinders with those made to an empty cylinder. If the two types of sniff are of similar size, the patient almost certainly has an impaired sense of smell.

Remedial Sniffing

We have glasses to help those with defective vision, hearing aids for the partly deaf, and who now will produce an artificial device to improve the smelling ability of people with subnormal noses?

—*Popular Science Monthly*, 1931

If perception and sniffing are inseparable, what happens to people who can't sniff? The most extreme case of nonsniffing is the person with a total laryngectomy, or removal of the voice box (larynx), a procedure that disconnects the upper and lower respiratory airways.

After laryngectomy, a person breathes through a hole in his throat, rather than through the mouth or nose, so he is unable to sniff or even activate his vocal cords to speak. Adding to their misery, about 85 percent of these patients are smell-impaired. Fortunately, some can be helped by a simple physical maneuver that resembles a polite yawn, or in other words, yawning with the mouth closed. This pseudo-sniff technique pulls air through the nose (though not the lungs) and allows about 50 percent of patients to score in the normal range on a smell test. A device called a tracheostomy valve, which directs exhaled air upward past the vocal cords and into the back of the nasal passages, restores speech function and also improves odor perception.

Impaired sniffing also occurs in Parkinson's disease and contributes to the smell loss found in these patients. Because the disease affects motor movement, the sniffs of a Parkinson's patient are weak and small. The worse their sniffing, the worse their performance on olfactory tests. The patients with the worst deficits can improve their test scores by simply taking bigger sniffs. While part of the problem lies in the physical action of the sniff, Parkinson's patients often develop cognitive impairment, which registers on smell tests; in fact, smell deficits are an early symptom of the disease.

A 1996 U.S. patent describes a device to help the sniff-impaired. It resembles a double-ended turkey baster, with the bulb in the middle equipped with one-way valves. The user positions one end of the device over, say, a bowl of chili, then squeezes and releases the bulb, and it fills with air. Now the user inserts the other end in his nostril and squeezes again, forcing a bulb full of chili-scented air up his nose. The device is sort of an Elsberg self-blaster, a nose trumpet for the hard of smelling.

Boosting nasal airflow even improves odor perception in normal people. The Breathe Right nasal dilator was first marketed in 1993 to help reduce snoring by increasing nasal airflow, but got attention as an athletic aid the following year when Herschel Walker of the Philadelphia Eagles wore one for the first time in an NFL game—he had a cold. When Jerry Rice of the San Francisco 49ers followed suit, the Breathe Right gained locker-room cred, and commercial success followed in drugstores across the country. The dilator is placed on the bridge of the nose just above the fleshy portion of the nostrils, where it exerts a springlike action that prevents the sides of the nasal vestibule from collapsing inward during an indrawn breath. (The nasal vestibule is the space behind the opening of the nostril; it's the finger-pickable part of the external nose.) Testing shows that wearing a dilator makes odors smell stronger, improves odor identification ability, and helps the wearer detect an odor at significantly lower concentrations. These benefits are the result of more air getting up into the nose. The nasal dilator increases the intensity of food aromas in the mouth but, weirdly, decreases the pleasantness.

THE ACT OF SNIFFING, overlooked by many scientists and politely ignored by well-mannered people, is critical to how we generate a mental image of the smellscape. The rapid sampling of odor-laden air is managed by a precisely timed interplay of sensory and motor function. In many instances, sniff improvement results in smell improvement. Seventy years after Charles Elsberg set out to suppress the sniff, we have finally begun to appreciate its value.

Even as it makes midsniff adjustments to the smell stream entering the nose, the brain is actively fine-tuning the mental impression it creates from an odor through a process called adaptation. Everyone is familiar with visual adaptation: after being in bright sunlight, it takes a minute or two for your eyes to adjust as you enter a darkened room. The reverse happens when you leave a movie theater in midday: the sunlight is unbearably bright at first, but gradually you adjust. Olfactory adaptation works on a similar principle: a new odor smells strong when we first experience it, but the longer we're exposed to it, the more it fades into the background. In the extreme, the smell may be undetectable for a while.

It's easy to overstate the practical importance of this phenomenon. Adaptation is a temporary change; it doesn't permanently erase the ability to smell. Fragrances are not written in disappearing ink: if women stopped smelling an eighty-five-dollar perfume within a few days of buying it, the fragrance industry would have collapsed long ago. The extent of adaptation depends on the nature of the smelling being done. Perfumers I know insist they can only smell half a dozen fragrances before they notice a dulling of perception. For these professionals, olfactory fatigue is a real obstacle. They sample trial perfumes from blotters, five-inch strips of filter paper dipped in the liquid. The professional takes a quick sniff or two and moves the blotter away, ever conscious of overdoing it.

In contrast, an amateur sniffer holds the blotter in front of his nose and inhales continuously, a sure-fire way to dull the nose. Even one minute of such deep breathing makes an odor immediately harder to detect. When I run a consumer smell test, I let the panelists sniff at their own natural pace. I've found they can easily assess a couple of dozen scents without a noticeable decline in performance. That's because they are sampling a variety of scents and doing so to make a quick thumbs-up or thumbs-down opinion— the typical objective of consumer and market research. This poses much less risk of adaptation than does the perfumer's repeated study of minor differences between related samples. The average person making rapid-fire judgments does not need to worry about the smellscape fading from view.

THE LONGER YOU are exposed to an odor, the more you adapt to it. Step into a garlic factory and the reek will overwhelm you. A few minutes later its intensity fades, and after an hour you might not be able to smell garlic at all, no matter how hard you try. Work there a few months and this adjustment will happen almost as soon as you step in the door. That was how I once became oblivious to *Safari*.

Early in my career, the company I worked for was developing the perfume for Ralph Lauren. As we tweaked the formula, ran stability tests, corrected the color, and did the million other chores needed to ensure a successful launch, the entire building was steeped in *Safari*. A few weeks into the job, none of us noticed it.

After a long vacation, I opened my closet to grab a suit for work, and got an overpowering faceful of *Safari*. The sensory truce between my nose and my workplace had fallen apart in less than two weeks. Similarly, long-term adaptation is what keeps plumbers and pig farmers from going insane.

Adaptation is a two-way street: when the odor source is removed, the nose gradually regains its sensitivity. This time-course of recovery is almost the mirror image of adaptation. Step outside after your visit to the garlic factory, and the recovery begins. If you were inside for just a few minutes, recovery will take a matter of minutes. If you were there for hours, it will be hours before full response returns. Odor strength is another factor in adaptation. The stronger the smell, the more you adapt. Ten minutes on the processing floor of the garlic factory will cause more adaptation than ten minutes talking to someone with garlic breath.

Adaptation is also odor-specific. If you work in a garlic factory, your nose will selectively tune out garlic, but your sensitivity to roses, sour milk, beer nuts, and other un-garlic-like smells will be unaffected. The narrowness of adaptation is sometimes exploited by perfumers when they try to match one fragrance to another. A perfumer will use saturation sniffing as the final step in comparing the target and the make. He sniffs the sample to the point of total adaptation, then smells the target; with his brain filtering out any sign of the original, any remaining minor differences will stand out.

Adaptation is a useful feature of any sensory system; it preserves our ability to detect small differences between stimuli against enormous variation in overall intensity. Just as auditory adaptation lets us have a whispered conversation but also talk in the middle of a rock concert, olfactory adaptation constantly recalibrates our noses to background conditions. It also selectively tunes new smells into the background, freeing our attention for the next new scent that may be creeping our way.

The Spin Doctors

In a lecture hall at the University of Wyoming in 1899, a chemistry professor named Edwin E. Slosson played a prank on one of his classes. He explained that he wanted to demonstrate the diffusion of odor through the air. He poured some liquid from a bottle onto a wad of cotton, making a show of keeping it away from his nose. He started a stopwatch and told the students to raise a hand as soon as they smelled something. Here's what he reports happened:

> While awaiting results I explained that I was quite sure that no one in the audience had ever smelled the chemical compound which I poured out, and expressed the hope that, while they might find the odor strong and peculiar, it would not be too disagreeable to anyone. In fifteen seconds most of those in the front row had raised their hands, and in forty seconds the "odor" had spread to the back of the hall, keeping a pretty regular "wave front" as it passed on. About three-fourths of the audience claimed to perceive the smell, the obstinate minority including more men than the average of the whole. More would probably have succumbed to the suggestion, but at the end of a minute I was obliged to stop the experiment, for some on the front seats were being unpleasantly affected and were about to leave the room.

Slosson's experiment vividly demonstrated the potency of olfactory suggestion, for he was holding a cotton ball soaked in nothing but water.

The sensory expert Michael O'Mahony revisited the phenomenon in the late 1970s. During a British television documentary on taste and smell, he showed viewers an electronic device that he claimed could capture and broadcast odors using "Raman Spectroscopy." The machine played a ten-second audio tone that viewers were told would evoke a "pleasant country smell." They were encouraged to call in or write and describe what they smelled. Many did. They reported smelling new-mown hay, freshly cut grass, lavender, honeysuckle, and so on. O'Mahony repeated the trick on a BBC radio show using a supposedly inaudible "ultra high frequency tone"—actually no sound at all. Some listeners reported smell sensations when it was played.

While amusing, these stunts by Slosson and O'Mahony raise serious questions for scientists conducting smell studies, because they show that just expecting a smell can trigger an odor perception. Thus a purely psychological expectation might have the same consequences as a real smell. For researchers the question becomes, How can we be sure the results of an odor experiment are really due to the smell and not to expectations about the smell? What is needed is an olfactory placebo: a test condition in which people are led to believe an odor is present when in fact it is not. To truly have an effect, an odor must outperform the placebo. This was the reasoning behind a study I did with Susan Knasko, a postdoctoral fellow of mine at the Monell Center, and the late John Sabini, a psychology professor at the University of Pennsylvania. We sprayed water mist in the air and told people it had a smell. The test room was actually scent-free and remained so. People who were told the smell was unpleasant later rated the room as smelling bad. When told the smell was pleasant, they liked the smell of the room. A supposedly "neutral smell" produced intermediate results. Interestingly, physical symptoms such as headache and itchy skin were also affected by the "good smell" and "bad smell." Our study was the first to confirm in the laboratory that the power of suggestion, by itself, could produce odorlike effects.

The psychologist Pamela Dalton and her colleagues took this

result and pushed it much further: they showed that expectations alter the perception of actual odors. She had volunteers sit in a test chamber for twenty minutes while exposed to odors that were neither pleasant nor unpleasant. Some subjects were told nothing about the odor. Others were told it was a potentially harmful industrial chemical or, alternately, that is was a distilled, pure natural extract. To use the Clinton-era term for expectation management, the experimental conditions differed only in spin. By the end of the test, all three groups had higher detection thresholds—their noses had been dulled by adaptation to the real odor. However, their perception of odor intensity was spin-dependent. With positive spin or no spin at all, the odor seemed less intense as time went on; with negative spin it smelled as strong or stronger. In other words, odors we think are benign fade from awareness, while those we believe to be hazardous hold our attention and stay strong.

It may not even matter whether the actual smell is good or bad. Spin can alter these perceptions as well. Dalton tested odors that were pleasant (wintergreen), unpleasant (butyl alcohol, a solventlike smell), and neutral (isobornyl acetate, a balsamlike note). Negative spin made all three smell stronger. Information bias is very effective at distorting the clear evidence of our senses—the brain easily trumps the nose.

Biasing information doesn't have to come from an authority figure in a lab coat. Dalton tested two people at a time in the environmental chamber. One was an unsuspecting volunteer, the other a carefully scripted actor pretending to be naive. The actor kept up an ongoing verbal and behavioral commentary about the odor in the air. This peer-to-peer kibitzing worked splendidly. When the spin was negative, 70 percent of volunteers reported health symptoms (everything from throat irritation to dizziness to stomachache); when it was positive, only 12 percent did so. Given a scent in the air—any scent—acquaintances can literally talk you into feeling sick.

The commonly acknowledged power of scent derives in large part from the power of suggestion. Negative placebo effects may exacerbate the symptoms of "sick building syndrome"—for example, if you believe that the musty smell in your office is from a toxic mold—while positive placebo effects explain the popularity of aromatherapy treatments. Beneficial mood change is one of the biggest claims made for aromatherapy. For example, lavender is usually extolled as relaxing and neroli as stimulating. A recent study showed that positive spin can completely reverse the aromatherapeutic effects of these two scents. When told the lavender they were smelling "has relaxing properties," people did in fact relax, as measured by changes in heart rate and skin conductance. Yet when told it "has stimulating properties," the same measures showed—presto change-o—that people were stimulated. The same reversal happened with neroli. It takes only the slightest waving of hands to create a positive placebo effect in aromatherapy.

The effects of spin often play out in everyday life. When the crew of a Norwegian air ambulance noted a cabbagelike smell in flight, they figured the patient they were transporting had passed gas and they ignored it. When the smell reappeared on another flight later that day, the crew was puzzled; it was unusual for two patients to be so extraordinarily gassy. Soon flames were shooting through the cockpit and the pilots were forced to make an emergency landing. The fartlike smell was smoldering insulation on electrical wires. The crew was in a medical mind-set, not a mechanical one, and their preexisting expectations led to a near-fatal misreading of what their noses were telling them.

Smells don't happen to a passive nose alone. The brain actively regulates the physical and cognitive aspects of odor perception: it exerts moment-by-moment control of sniffing to govern how much scent enters the nose; it systematically dials down the intensity of one smell to prepare us for the next; it automatically makes a provisional interpretation of a smell, based on context cues, to prime us for a behavioral response. From sniff to spin, the nose and brain constantly reshape our awareness of the smellscape.

CHAPTER 5

A Nose for the Mouth

Blindfold a person and make him clasp his nose tightly, then put into his mouth successively small pieces of beef, mutton, veal, and pork, and it is safe to predict that he will not be able to tell one morsel from another. The same results will be obtained with chicken, turkey, and duck; with pieces of almond, walnut, and hazelnut. . . .

—HENRY THEOPHILUS FINCK (1886)

WHEN IT COMES TO FOOD, I'M A SMELL CHAUVINIST: taste is boring. The tongue supplies just five channels of information: bitter, sweet, sour, salty, and umami. (My Japanese colleagues insisted for years that monosodium glutamate delivered more than a salty impression. The discovery in 1996 of glutamate receptors on the tongue finally proved their case. The savory taste of umami is now in the official pantheon.) While five taste channels are nothing to sneeze at, they're rudimentary compared with the 350 different receptors and two dozen perceptual categories available to olfaction.

There is another reason why I think taste is overrated. We are accustomed to experiencing flavor as a singular sensation in the mouth. As a result, we use the words "taste" and "flavor" interchangeably in casual conversation. This makes it easy to forget that flavor is actually a fusion of taste and smell, and that the apparent simplicity of flavor is just an illusion, one that is somtimes reinforced by language. For example, there is only one word for taste and flavor in Spanish (*sabor*), German (*geschmack*), and Chinese (*wei*). I think the tongue gets more credit that it deserves.

That smell makes the far greater contribution to flavor becomes obvious once it is taken out of play. Pinch shut the nostrils, and flavor disappears. What's left, as the American philosopher and critic Henry T. Finck noted 120 years ago, is bland texture. Caviar tastes like salty oatmeal, and coffee is merely bitter water. This simple, powerful truth is ignored by those who claim the sense of smell is weak and of little importance to modern humans. For example, the pop-science icon Carl Sagan once said "it is clear that smell plays a very minor role in our everyday lives." *Science Digest* claimed, "Modern man seldom uses the sense of smell except to detect a burning roast in the oven, or to enjoy a rose bush." The pioneer sexologist Havelock Ellis had such contempt for smell that he tried to minimize its role in flavor: "If the sense of smell were abolished altogether the life of mankind would continue as before, with little or no sensible modification, though the pleasures of life, and especially of eating and drinking, would be to some extent diminished." One hesitates to imagine what sort of cramped, joyless inner life could lead a person to write such things, for the reality, made clear by Finck's demonstration, is that the sense of smell contributes mightily to our enjoyment of food and for this alone deserves to be celebrated.

In his essay on "The Gastronomic Value of Odours," Finck described a particular type of smelling we use to savor food. He pointed out that aromas released from food in the mouth reach the nasal passages via the back of the throat, and are exhaled through the nostrils. The act of swallowing drives aromas along this reverse path. In effect, we smell our food from the inside out. Today this is known as retronasal olfaction, but I prefer Henry Finck's name for it: a "second way of smelling," a phrase that sets it apart from the usual nostrils-first mode. Retronasal olfaction has become a hot topic among sensory scientists, and recent findings confirm Finck's intuition: the second way of smelling operates by its own set of sensory rules.

THE TWO PHYSICAL paths to the nose—one from the outside world and the other from the mouth—have parallels in the psychology of odor perception. The apparent location of a smell—inside or outside of our body—determines how we perceive it. The psychologist Paul Rozin demonstrated this in a simple experiment. He taught people to recognize the smell of four unusual fruit juices. They sniffed the samples while blindfolded, and quickly learned to tell the them apart with perfect accuracy. When Rozin squirted the same juice samples into their mouths with a syringe, they could not identify them reliably. A smell well-learned when sniffed by the nose is poorly recognized in the mouth. This suggested to Rozin that location has consequences: a food smells one way "out there" and a different way "in here." The psychological difference between outside-in and inside-out smelling, when combined with taste sensation from the tongue, produces strange contrasts. It makes for foods that smell good but taste bad (coffee, for example), and others that smell bad but taste good (blue cheese).

The psychologist Debra Zellner studies a peculiar sensory illusion involving sight and smell. She pours a clear, scented liquid into two glasses and adds color to one. To a blindfolded person the two samples smell equally strong; with the blindfold removed, the colored version smells stronger. In the classical version of this color-odor illusion, the liquid is sniffed by nose. Zellner wondered what would happen if the smell were delivered by mouth. She had people sip the samples through a straw—the liquid was visible under a clear plastic lid, which prevented through-the-nostril smelling. Under these conditions the illusion was reversed: adding color reduced perceived odor strength.

Because smell and taste are inextricably linked in flavor perception, experience in one modality can affect the other. For example, some odors are commonly described in terms of taste: honey smells "sweet" and vinegar smells "sour." The Australian psychologist R. J. Stevenson and others have shown that odors acquire taste qualities through associative learning. After a novel odor is paired a few times with the sweet taste of sucrose, the odor is perceived as smelling sweet. If paired with citric acid, it seems to smell sour. This cross-sensory link works in the other direction as well: smells can alter tastes. Strawberry odor, for example makes a weak sugar solution taste sweeter, and a whiff of soy sauce boosts the perceived saltiness of a saline solution. Sensory researchers have just begun to understand the psychological interplay between smell and taste. They are now looking at how these senses are neurologically cross-wired in the brain. To a smell-centric guy like me, the study of taste is about to become much more interesting.

The Pleistocene Barbecue

Carnivores rarely savor their food: they rip, chomp, and swallow. Herbivores chew for hours on end, not for sensory pleasure but to make tough, fibrous plant matter digestible. Humans, in contrast, anticipate, savor, and linger over the aroma of food. We go to great lengths to increase the appeal of food by cooking it and adding spices. The second way of smelling not only provides the pleasure we take in eating, but also may be the key to how the human sense of smell has evolved over time.

Traditionally, researchers in cultural anthropology and sociology have treated food preparation as an expression of culture, as a collection of behaviors driven by custom and creativity only. A new generation of behaviorally oriented evolutionists is now challenging this profoundly unbiological point of view. The Harvard University anthropologist Richard Wrangham, for example, sees cooking not as an optional behavior—a cultural frill—but as a biological require-ment for human survival. Surveying the evidence, he finds that "no human populations are known to have lived without regular access to cooked food." Even the Inuit hunters of the Arctic, famous for their raw diet, occasionally cooked their whale blubber.

Hominids—the near-human species that link us to our common ancestor with the chimps—were definitely cooking with fire 250,000 years ago. Wrangham finds evidence of cooking as far back as 790,000 years, and speculates that it may have begun as far back as 1.7 million years ago. In any case, cooking with fire was well estab-lished when our first anatomically modern ancestors emerged in Africa some 100,000 years ago. We've grilled a lot of mastodon steaks through the ages.

The invention of cooking had profound consequences for diet and social behavior. Cooking releases nutrients and makes vegeta-bles faster to eat and easier to digest. Wrangham calculates that for a 120-pound woman to take in 2,000 calories a day, she would have to eat eleven pounds of raw fruits and vegetables. That's a lot of time at the salad bar. Clinical studies show that raw-food cultists in Ger-many struggle to keep up nutritionally with their countrymen: they suffer from chronic energy deficiency and the women fail to men-struate. If European sophisticates with desk jobs and handy super-markets can't thrive on a raw-vegetable diet, how well would a band of hunter-gatherers do?

Adding meat greatly enhances the diet. Chimpanzees in the wild are big fans of monkey meat, but even with their powerful jaws they take hours to gnaw the raw flesh from a bone. Given the effort involved, raw meat isn't a routine source of nutrition for chimps. Nor would it have been for early hominids. A *Homo erectus* female (our evolutionary cousin) would have needed six hours a day to get all her calories from raw meat, according to Wrangham's calculations. Cooked meat, however, is a different story: it is nutrient-dense, easily chewed, and rapidly consumed. The time saved by cooking changes our behavior patterns. Where all other large primates snack throughout the day on raw fruits and leaves, we eat a few discrete meals, leaving more time for other activities. The widespread popularity of cooking among protohumans meant that powerful jaw muscles and large teeth were no longer essential, and as their evolutionary advantage shrank, so did they. In the last 100,000 years our teeth and jaw muscles have become even smaller, making possible finely controlled chewing movements of the tongue and jaw. The more nimble modern mouth makes an easy-to-swallow "bolus" of food and releases more aroma in the process. In the long run, cooking has literally changed the shape of our face.

COOKING HAS ALSO changed our sensory world: it introduced novel aroma molecules and whole new classes of smells. The savory notes of roasted meat, toasted nuts, and carmelized vegetables were rare accidents before we fired up the Pleistocene barbecue. More new smells—baked bread and boiled mush—arose with the cultivation of wheat and other grains about 12,500 years ago. Sheep, goats, pigs, and cattle were all domesticated roughly 10,000 years ago. With them came the smell of butter and the fermented bouquets of yogurt and cheese. As early villagers mastered the art of fermentation, the heady aromas of beer and wine joined the mix.

We are a cooking species, and the smell of an impending meal is woven into our biology. Food aroma is an invitation and a spur to action. Even before the first bite, it triggers an elaborate sequence of physiological events: salivation, insulin release by the pancreas, and the secretion of various digestive juices. The aroma of bacon, at a level so faint it can't be consciously identified, has been shown to trigger the flow of saliva. This would not have surprised cookbook author James Beard, who once said, "Nothing is quite as intoxicating as the smell of bacon frying in the morning, save perhaps the smell of coffee brewing." We expect to be stimulated en route to a meal— the anticipatory smells of cooking have become almost a biological requirement. This is a big headache for manufacturers of prepared foods. The physics of microwave heating doesn't create the toasted, roasted, and caramelized notes that signal impending "doneness." Food companies spend a lot of time and money on technological work-arounds to restore these missing scents.

IN ADDITION TO cooking food, we spice it. Spice use is a universal human habit, though there are significant regional differences in the spices that are used and how they are combined. What qualifies as a spice? In one definition, it's "any dried, fragrant, aromatic, or pungent vegetable or plant substance, in the whole, broken, or ground form, that contributes to flavor, whose primary function in food is seasoning rather than nutrition, and that may contribute relish or piquancy to foods or beverages." Roots, seeds, dried leaves, even aromatic lichens fit this definition; including fresh herbs adds still more materials. There are a lot of spices, and yet, like the huge number of possible smells in the world, the closer one looks, the more this apparent diversity can be simplified. At the core of each of the world's great culinary traditions is a small group of spices and flavorings. A perfumer would think of these combinations as an accord, the key ingredients that define a style of perfume. The late food expert Elisabeth Rozin called these combinations "flavor principles": "Every culture tends to combine a small number of flavoring ingredients so frequently and so consistently that they become definitive of that particular cuisine." Rozin could conjure up an entire culture using two or three key flavorings. She rarely had to use more than four. For example, soy sauce, rice wine, and gingerroot form the Chinese flavor principle, while the Hungarian one consists of paprika, lard, and onions. A beloved and easily recognized flavor principle gives ethnic authenticity to whatever is cooked in it. In the future,

Hungarian deep-space explorers eating processed algae paste will find it quite palatable as long as it is seasoned with paprika, lard, and onions.

Some spices are used by many different cultures. What makes a flavor principle distinctive is its *specific combination* of seasonings. Consider lemon, a widely used flavor source. Add cinnamon, oregano, and tomato and you've got a Greek principle. Add fish sauce and chili and you've got Vietnamese. The extensive overlap in ingredients across flavor principles means that every traditional cuisine on the planet can be prepared from a very short shopping list. The thirty or so principles Rozin describes in her book require about four dozen ingredients. All the flavors of world food culture can fit into a single grocery bag.

Liz Rozin's theory of food aroma strikes some people as too minimalist to account for the richness of human cuisine. What they fail to appreciate is the power of combinatorics, which makes it possible to generate huge numbers of flavor variations from a few basic odorous elements. The Chicago chef and restaurateur Charlie Trotter understands this. "You can prepare forty dishes from six ingredients," says Trotter. He likens creative cooking to jazz improvisation. A chef who has mastered the basic repertoire—the classical flavor combinations—can improvise endless new dishes with only a handful of spices. Thus the cook and the chemist have arrived at the same fundamental truth: sensory diversity is achieved with relatively few ingredients. The chemist can re-create the aroma of any foodstuff with fewer than a thousand odor molecules, and the chef can build any global cuisine with a few dozen spices. The amazing variety of human cuisine, at the chemical as well as the aesthetic level, is a matter of basic themes and endless variations.

THE CORNELL UNIVERSITY evolutionary biologist Paul Sherman is another scientist rethinking the assumption that all variation in food habits is cultural. Sherman studies how spice use relates to human survival. He and his collaborator Jennifer Billing were intrigued by the fact that spices often have antimicrobial properties: they contain natural chemicals that kill bacteria and fungi. Could the point of cooking with spices be to reduce spoilage and food-related illness? To test their idea, Sherman and Billing assembled a collection of ninety-three cookbooks from thirty-six countries. From these, they selected 4,578 meat-based recipes and meticulously noted what spices were used in each.

On a worldwide basis, nearly every meat dish (93 percent) had one or more spices. The results varied, however, with a country's climate: the number of spices per recipe increased with the average annual temperature. In Finland and Norway, for example, one-third of recipes used no spices at all. In contrast, in Ethiopia, Kenya, Greece, India, and Thailand, every recipe called for at least one spice. Sherman and Billing ran other statistical analyses and found that average annual temperature was correlated with the proportion of recipes containing spices, and the total number of spices used. Since unrefrigerated meat goes bad faster in a warm climate, more spices might mean better protection against spoilage. Sherman and Billing examined the antibacterial power of the various spices, and found that the hotter a country, the more bacteria species are inhibited by the local selection of seasonings. They conclude that while spice use is something we do because it tastes good, it also rids food of pathogens and therefore provides a biological advantage in keeping people healthy. (They briefly considered whether spices might be used to mask the bad taste of spoiled food, but dismissed the idea as a nonstarter: there would be little benefit to survival in encouraging people to ingest toxins.)

In their tally of thousands of meat-based recipes, Sherman and Billing found that the most commonly used spices are onion (in 65 percent of all recipes) and pepper (63 percent), followed by garlic (35 percent), hot peppers (24 percent), lemon and lime juice

(23 percent), parsley (22 percent), ginger (16 percent), and bay leaf (13 percent). Another thirty-five spices appear only occasionally (in 10 percent or fewer of all recipes). They found that the vast majority of the world's recipes could be made with about four dozen spices—a number remarkably close to the length of Elisabeth Rozin's world cuisine shopping list. Further, the average meat recipe calls for 3.9 spices, a number that is consistent with Rozin's flavor-principle concept.

Sherman returned to his cookbook collection and analyzed another 2,129 recipes, this time looking at only vegetable dishes. Compared with meat recipes, these use fewer spices (2.4 per recipe on average). Still, the results supported the antimicrobial hypothesis: the hotter the climate, the more spices, though this relationship proved somewhat weaker for vegetable dishes. Why? Fruits and vegetables come prepackaged with physical and chemical defenses against microorganisms, which makes the health benefit of adding spices correspondingly smaller.

Cooking and spicing are behavioral adaptations with biological consequences. They have shaped our face and made mouth-based smelling a defining human trait. Outlandish as it sounds, spicy cooking may even have altered the core of our biological identity—our DNA.

It is often said that a species's DNA code can be read like a book. If so, some biologists read it like the sports pages—they add up the number of odor receptor genes and rank us against other species according to the results. Rats lead the Mammalian League with the most functioning receptor genes; dogs and mice are a few games behind, while chimpanzees and humans are looking for a wild-card berth; and dolphins—an aquatic expansion team—own the cellar.

Among primates, humans have the highest proportion of non-functional receptor genes; we keep a lot of obsolete junk in our genetic attic. Superficially, it looks like the human nose is weak (relatively few receptors) and getting weaker (losing receptor genes at four times the evolutionary rate of other higher primates). Some, such as the science writer Nicholas Wade, see this as a case of use it or lose it. He says that "the price of civilization is that the faculty of smell is inexorably being degraded." Wade's gloomy conclusion may not be justified. Humans continue to evolve, and geneticists have identified hot spots in our genome—areas of biological function in which new genes are being born. Olfaction is one such hot spot. In the last 5,000 to 10,000 years, genes for smell receptors, along with genes related to diet and metabolism, have been evolving faster than those in any other physiological system. One new study finds that "many changes in the human olfactory repertoire may have occurred very recently," the changes in this case being beneficial genetic mutations that have become fixed traits throughout the population.

The human genome responds rapidly to cultural changes. For example, in ancient populations the gene for lactose absorption ceased to function shortly after weaning. With the emergence of dairy farming, natural selection favored individuals in whom the gene stayed active into adulthood. The selective advantage of being able to eat milk products was so great that adult lactose absorption became a widespread trait within 5,000 years, a mere blink of the eye in evolutionary time. I suspect there has been ample time for the aromas of cooked food to influence our odor receptor repertoire in a similar way. If our gut evolved to digest dairy products, why wouldn't our nose evolve to appreciate the smells of cheese, butter, and yogurt?

In the recent evolutionary past we have evolved entire subfamiles of odor receptors not shared by the chimpanzee—our closest living relative. An intriguing possibility is that these new receptors are tuned to detect new smells—ones that only recently became important to human survival. It's speculation on my part, but I'd bet these receptors pick up the nuances of grilled meat—salmon filets and mastodon steaks—along with the volatiles of fermentation: not only milk products, but alcoholic drinks from beer to wine. On a daily basis we season food to please our palate, but over the long run our palate is evolving to match our menu.

I also suspect that dogs are part of the whole story. Dogs were first domesticated by man somewhere in Siberia about 15,000 years ago, just as human populations were shifting from a hunter-gatherer existence to sedentary village life. Increasingly preoccupied with the complex man-made aromas of the cooking pot, our ancestors began to rely on hunting dogs to locate the telltale scent of game. Having co-opted the canine nose, our own scent-tracking ability began to fade. Dogs became, in effect, our long-distance noses, while we specialized in the close-in smelling of food in the mouth.

Dog and humans have complementary nose skills: dogs have little retronasal ability but great distance detection; humans vice versa. (I'm unable to find a single scientific paper on canine retronasal smell. According to pet-food manufacturers, dogs sniff first and gulp later; they don't spend a whole lot of time savoring food in the mouth.) The Yale University neurobiologist Gordon Shepherd suggests that retronasal smelling "has delivered a richer repertoire of smells in humans than in nonhuman primates and other mammals." I would go further and claim that humans are a retronasal species; our best olfactory skills are reserved for appreciating food aromas at the point of eating. Our talent is smelling food in the mouth, not food on the hoof. When it comes to tracking the scent of a gazelle on the savannah, we can't compete with our hounds; but once we drag it back to the campfire we can sure season the hell out of it.

CULTURES ALL OVER the world may choose from the same selection of spices, but that doesn't guarantee that we all find each other's cuisines equally appealing. Aromas mark differences between cultures, along with all the moral baggage that entails. On a field trip to Costa Rica, when Miss Stevens admonishes him to "respect other cultures this instant!" Eric Cartman replies, "I wasn't saying anything about their culture, I'm just saying their city smells like ass." Offhand dismissals of cultural differences aren't limited to the fourth graders of *South Park*. Before he became the president of France, Jacques Chirac was mayor of Paris, and made himself notorious for observing that "the noise and the odor" of freeloading immigrant families would reasonably push a hardworking Frenchman over the edge. He hastened to add, in Cartman fashion, "It is not racist to say this."

Smell prejudice is not just a Eurocentric trait. Wang Lung, the fictional hero of Pearl Buck's *The Good Earth*, moves to another region of China where his scent marks him as a outsider: "[W]hen an honest man came by smelling of yesterday's garlic, they lifted their noses and cried out, 'Now here is a reeking, pig-tailed northerner.' The smell of the garlic would make the very shopkeepers in the cloth shops raise the price of blue cotton cloth as they might raise the price for a foreigner."

Anthropologists tell us that olfactory stereotyping is central to tribal identity. The Desana people of Colombia's Amazonian rain forest, for example, believe each tribe has a characteristic odor due partly to heredity and partly to what it eats: "Thus, the Desana, who are hunters, are said to exude the musky smell of the game which they eat. Their neighbours, the Tapuya, on the other hand, live by fishing and are thought to smell of fish. The nearby Tukano are agriculturalists and they, in turn, are said to smell of the roots, tubers and vegetables which they grow in their fields." Traditional Scottish clans put a different spin on it. Before the invention of woven tartans, each clan was associated with a plant, worn by its members as an aromatic badge of identity. An enterprising smell scientist is attempting to reintroduce the concept by marketing clan-based perfumes. Eau de Whortleberry, anyone?

At cultural boundaries the smell of food become an invisible, fragrant fence. One study went to the trouble to prove that bonito flakes smell like food to Japanese people but not to Germans; the opposite is true for marzipan. You eat what you were raised on. The most unsettling result of this study was that nearly 40 percent of the German ladies interviewed found the smell of Vicks VapoRub to be edible.

Does the fragrant fence limit us to the food aromas of our birth culture? Not necessarily; but there are hazards in jumping the fence. These are nicely depicted in Radhika Jha's novel *Smell*. Leela, a young Indian woman born in Kenya, is sent to live with relatives who run an Indian grocery in Paris. Aromatic crosscurrents are present from the opening sentence: "When the wind blew hard, as it did very often that spring, the smell of fresh baguette would fight its way into the Epicerie Madras to do battle with the prickly smell of pickles and masalas." Leela has a fine awareness of scent and is skilled at cooking with traditional Indian spices. As she learns the ways of Paris she improvises new dishes and creates new possibilities for her love life and career. (She takes a French lover and becomes the darling of the Parisian fusion cuisine scene.) Eventually, Leela realizes that the scents that make her exotic and attractive also make her an outsider. As an author, Radhika Jha has an extraordinary feel for the boundary-creating power of scent, perhaps because she herself lived in Paris as an exchange student. By showing how one woman used scent to redefine her relation to two cultures, she proves it is possible to cross the fragrant fence.

SOME FOOD AROMAS raise the fence to unscalable heights. For example, if you are not Swedish it is unlikely that you can be persuaded to try Surströmming. Surströmming is fermented herring, and is horrifically foul-smelling even to those who consider it a national delicacy. Another Scandinavian specialty is lutefisk. To make it, one soaks air-dried codfish in water for several days, then in a solution of caustic lye for another couple of days, and ends with a few more days in plain water. The result is a swollen, jellylike mass of smelly fish flesh that is popular in Norway and the Norwegian-heavy precincts of Minnesota and Wisconsin. Garrison Keillor recalls lutefisk as "a repulsive gelatinous fishlike dish that tasted of soap and gave off an odor that would gag a goat." But people who consider themselves true Sons of Knut eat it at least once a year. Norwegians are not insane; they know lutefisk smells bad. But they have carved out a special exemption for it—they've made it a badge of belonging.

The psychologist Donald E. Brown compiled a list of cultural universals that includes things like music, proverbs, incest avoidance, and death rituals. I would like to propose an addition to the list of universals: every culture has a foul-smelling food for membership. You are not really Taiwanese unless you eat "stinky tofu" (chunks of fermented soybean curd). You are not really Icelandic unless you eat harkarl (rotten shark meat). Real Japanese eat natto (a gluey mass of fermented soybeans that smells like creosote). Then there is the fabulously stinky durian, or jackfruit, of southeast Asia. Singapore being Singapore, one is allowed to eat its sweet, custardy innards, but it is illegal to carry it on public transportation. I'm personally a big fan of kimchi, the national condiment of Korea. It's made from fermented Chinese cabbage, vinegar, garlic, fish sauce, and lots of red pepper. It packs a punch—a bottle of it once exploded in my refrigerator. Its postingestive consequences are spectacular: the humorist P. J. O'Rourke described them as "a miasma of eyeglass-fogging kimchi breath, throat-searing kimchi burps, and terrible, pants-splitting kimchi farts."

AMERICA IS IN the midst of a great sensory reawakening; we are more open to new foods and flavors and smells than at any point in our history. In a country where quiche was once considered exotic, we are no longer surprised to find pad thai in Peoria or moussaka in Muskogee. Kraft Foods, an outfit best known for serving up millions of pounds of macaroni and cheese, recently introduced a Mango Chipotle seafood marinade. Yet in contrast to this growing abundance of sensory options, the regional differences that once characterized the national smellscape are fading. In 1947 *The Saturday Evening Post* asserted confidently that "West Coast doughnut flour has a predominant lemon flavor, whereas in New England, doughnuts have a strong nutmeg flavoring, with little lemon." Traces of these regional preferences linger on the contemporary American scene, as evidenced by variation in air-freshener sales. Food-inspired scents such as vanilla and cinnamon have a 39-percent market share in the North Central states, compared to only 28 to 29 percent in the Northeastern, Western, and Southern regions. Citrus and fruity scents (lemon, orange, grapefruit, mandarin, and green apple) show the reverse pattern: they are only 16 percent of sales in the North Central states, but 22 to 23 percent elsewhere.

Beer brewing used to be a strictly local operation, but today's American beer market is dominated by national brands like Miller and Budweiser. It is no coincidence that in the last fifty years the amount of malt in the average American brew has declined more than 25 percent and the amount of hops more than 50 percent. In other words, beer is less bitter and less aromatic than ever. Big brands expanded through a strategy of making inoffensive beer. They traded character for market share. The good news is that the microbrewery movement is thriving. Small producers have created distinctive beers with greater flavor and more interesting aromas. These so-called "craft" beers are on the rise, while overall domestic beer sales are flat or declining.

It is tempting to think of odor blanding as a typical expression of American mass-market consumerism. Yet it's a truly global phenomenon. France, a country not known for its welcoming attitude toward American culture, is home to some of the world's stinkiest cheeses: St.-Nectaire, Ami du Chambertin, and Epoisses. (The last is said to smell simultaneously of "socks, wet dog, and manure.") France has more varieties of cheese on sale today than ever before; roughly a hundred new varieties hit the market annually. Paradoxically, these products are tasting more and more alike. Traditional mold-ripened cheeses, made from unpasteurized milk, change in texture, smell, and taste as they age. The new versions are made from pasteurized and ultrafiltered milk; they are built for a long and consistent shelf life. Industrial brie—rubbery, flavorless, and never-aging— is taking over the market.

French manufacturers go to great lengths to create an aura of authenticity for the new fromage-blah; they package it in a wooden box, wrap it in plastic straw, and give it an impressively historical name. In trade jargon, these impostors are known as *vrai-faux produit traditionnel*; think of it as cheese that is "fake but accurate."

Coffee Beans and Other Bad Memes

> Joel Lloyd Bellenson places a little ceramic bowl in front of me and lifts its lid. "Before we begin," he says, "you need to clear your nasal palate." I peer into the bowl. "Coffee beans," explains Bellenson's partner, Dexster Smith. "This is what they use in perfume stores. It's like the reset button." Dutifully, I reinitialize my nose by sniffing the beans.
>
> —CHARLES PLATT, *Wired* magazine, 1999

Charles Platt began his *Wired* cover story with this vignette about the two founders of DigiScents, Inc. Joel and Dexster had come up with a small unit that could release innumerable combinations of scent when activated by a digital signal from a personal computer.

Stanford graduates, with degrees in bioscience and engineering, respectively, they had previously started a successful genomics company. Neither of them knew beans about smell. That's why I had been hired a few months earlier—to bring a working knowledge of sensory science and the fragrance industry to the new venture. I thought their coffee stunt was silly. I'd seen beans at a trade show, but had never heard of a perfumer using them. Still, Joel and Dexster had an unerring sense of publicity—a useful talent for founders of a Silicon Valley startup. So I sat back and watched with inward eye-rolling as the meme of a "reset button for your nose" was launched into digital culture.

The bean meme is now a fixture in perfume retailing. I toured the Mall at Short Hills, New Jersey, recently and marveled at how thoroughly it has taken root. At the *Angel* counter in Nordstrom a glass cone full of coffee beans was held aloft on a brushed metal stand. In Bloomingdale's the beans were in a cocktail glass. The Jo Malone display in Saks had them in an apothecary jar with a metal lid. It's all good fun and marketing, but there is not a jot of science behind it. (There are twenty-seven aroma impact molecules in roasted arabica coffee—how could smelling all these help clear the nose?) I don't make an issue of it when I'm shopping, but a perfumer of my acquaintance was once ejected from a Nordstrom in Seattle for disputing the bean meme a little too persistently with the lady behind the counter.

The idea of a reset button for the nose goes back a long way. At nineteenth-century Japanese incense parties (which were part guessing game and part poetry contest), it was customary for participants to rinse occasionally with a mouthful of vinegar to keep the sense of smell sharp. American perfumers in the 1920s sniffed camphor to restore sensitivity after a hard day at the office. The pioneering odor classifiers E. C. Crocker and L. F. Henderson routinely sniffed camphor or ammonia to "refresh the nose" during long smelling sessions. It's not clear if these practices worked as intended, or if they are just testimony to the olfactory placebo effect. Similarly, contemporary food companies require taste-test panelists to rinse between samples. The rationale—that it minimizes flavor carryover—seems so commonsensical that no one bothered to test it until 2002. When a sensory lab finally got around to it, the results were surprising. In the study, trained tasters rated the bitterness of cream cheese samples mixed with different amounts of caffeine. (Caffeine is notorious for the delayed onset and lingering aftertaste of its bitterness.) Between samples the tasters tried all sorts of lab-standard palate-cleansing techniques: they rinsed with water or with sparkling water (up to six times); they ate carrots or crackers or plain cream cheese. The results were all the same—cleansing the palate made no difference to subsequent judgments of bitterness. Caffeine leaves a bitter taste, but panelists can compensate for it as they move from sample to sample. So go ahead—serve bread and crackers at your wine tasting, and enjoy the between-course sorbet at your fancy French restaurant. Just don't expect either habit to make your palate sharper.

According to discriminating foodies, red wine should be paired with only certain kinds of cheese. Aged Gouda, Dry Jack, and Manchego enhance the flavor of red wine, while blue cheese and triple cream varieties interfere with it. At least that's the dogma. Like many rules of cuisine, the logic behind wine and cheese pairings has seldom been put to a scientific test. The sensory specialist Hildegaarde Heymann and a graduate student addressed the question head-on. They trained panelists to rate red wines along a number of sensory dimensions. When wines were paired with eight different cheeses, the tasters' perception did indeed change, but not for the better. The flavor of the cheese accentuated the butteriness of the wines, but it blunted every other sensory characteristic— probably not what one wants when uncorking a valuable vintage.

Wine-tasting tradition holds that a wine must be drunk from the correct glass: reds from a large, bulb-shaped one that tapers at the mouth, whites from a smaller version of this, or perhaps from one that isn't tapered. The idea is that the size and shape of the glass determine how the aroma is collected and delivered to the nose, and that there is an optimal glass for each type of wine. Do these rules have a basis in fact, or are they simply the pretension of wine snobs? Only three studies have addressed the question, and the results are mixed. In one, a Mondavi cabernet smelled less intense in the traditional big-bulb Bordeaux glass than in other shapes; other sensory measures (fruitiness, oakiness, etc.) were unaffected by glass shape. Another study served red and white wines in five different glasses and found that shape altered the perception of the wines on nearly every rating scale. Why such different results? For one thing, the first study was done with blindfolded subjects and the second one was not. A judge's expectations about the wine change when the glass can be seen. A third study found that tapered, bulb-shaped glasses produced a stronger impression of wine aroma than a tulip-shaped or a nontapered bulb. This effect disappeared, however, when the odor sensitivity of individual judges was taken into account. Only people with superior noses could appreciate the subtle effects of glass shape. While this will no doubt reinforce the self-regard of wine snobs, the final joke is on them. The study presented a single wine in glasses of various shapes; afterward most judges thought they had been served two or three different wines. Another triumph of the visual over the aromatic. In the final analysis, glass preferences may be nothing more than a tradition. In a similar way, I have heard French perfumers insist that their style of smelling blotter (folded lengthwise into a V-shape, and cut to a point on the end) is superior to the thin, rectangular version used by Americans. Why? Because it allows the perfume to evaporate more precisely. The world of olfaction is filled with irrational beliefs, and sometimes that's just part of the fun.

CHAPTER 6

The Malevolence of Malodor

And when euyl substance shall putrifie,
Horrible odour is gendred therbye;
As of dragons & men that long dede be,
Theire stynche may cause grete mortalite.

—THOMAS NORTON, *Ordinall of Alchimy*
(late fourteenth century)

ON A LATE-SEPTEMBER SUNDAY IN 1971, I WALKED along a dusty footpath toward some oak woods near San Rafael, California. I was with a few oddly dressed friends: the men wore tights and jerkins, the women long-sleeved, flowing dresses and conical hats. I wore a Puritan robe with white collar and carried a wooden recorder. We were in a long line of costumed people stretching from a field of parked cars to the crest of a hill, where flew the pennants of the Renaissance Pleasure Faire. The wooded hills of Marin County were a congenial spot for this deliberate flight of fancy into the past.

Amid the bawdy puppet shows and the racket of tambours and sackbuts, one could almost slip into the mental habits of an earlier time. In Elizabethan England, bad-smelling air was thought to be the cause of disease. In *Hamlet*, Shakespeare wrote: "'Tis now the very witching time of night / When churchyards yawn, and hell itself breathes out / Contagion to this world." According to Simon Kellwaye in *A Defensative Against the Plague*, written in 1593, illness results from "some stinking doonghills, filthie and standing pooles of water and unsavery smelles." In a time before indoor plumbing, when open sewers were the norm, there were enough "stinking doonghills" to make everyone feel threatened by disease. For Elizabethans, however, odor was both a cause and a cure. They believed that good odors could ward off disease. This led them to hang spice-filled pomanders from necklaces and to fumigate their houses by burning incense, sulfur, and gunpowder. Beneficial aromas were so sought-after in times of plague that price-gouging was common. A writer in 1603 complained that rosemary, "which had wont to be sold for 12 pence an armefull, went now for six shillings a handfull."

Twenty years after the Pleasure Faire, people in Marin County were once more channeling a Medieval mind-set, and this time it wasn't fun and games. Like a pitchfork-wielding rabble demanding protection from plague-inducing vapors, antifragrance activists were out to ban perfume because they believed it was making them sick. They objected not just to perfume, but to the lingering scent of shampoo, body lotion, hair spray, deodorant, laundry detergent, and fabric softener. Protesting at a perfume industry meeting in San Francisco, activists wore respirators and carried a prop barrel labeled CALVIN KLEIN and TOXIC CHEMICALS. The "disability coordinator" for the San Francisco mayor's office joined the fray. "Ten years from now it will be politically incorrect to wear perfumes in public," he proclaimed. Even by the flamboyant standards of the Bay Area, this was great political theatre. But it raised an important question: Can a smell actually make us sick?

THE PROTESTERS were people who suffered from what they called Multiple Chemical Sensitivity, or MCS. They claimed to be so sensitive to chemicals in perfume that the slightest whiff would trigger symptoms. I spoke with several MCS patients at the time, and was struck by how unhappy and miserable they were. Their extreme efforts to avoid scented people and smelly places made them virtual shut-ins. One woman had moved her family to the Arizona desert in the hope that living in an isolated trailer custom-built with "nontoxic" metal and tile surfaces would solve her problem. It didn't. It was clear to me that these folks were genuinely distressed and deserving of sympathy. What wasn't clear to me was the nature of their illness.

Despite numerous investigations by medical experts and public health authorities, including the World Health Organization, there is no precise definition of MCS. According to a paper in *Occupational and Environmental Medicine*, it is "a poorly understood and controversial syndrome. Common symptoms include fatigue, difficulty concentrating, pounding heart, shortness of breath, anxiety, headache, and muscle tension. They occur 'in response to demonstrable exposure to many chemically unrelated compounds at doses far below those established in the general population to cause harmful effects. No single widely accepted test of physiological function can be shown to correlate with symptoms.'" The American Medical Association looked into MCS and decided in 1991 not to recognize it as an official diagnosis. In the meantime, MCS has been renamed Idiopathic Environmental Intolerance (or IEI) to reflect the fact that it has no known cause (i.e., it is idiopathic).

Amid all this confusion, IEI patients are consistent about one thing: they claim to be far more sensitive to odors than are other people. This is an easily testable proposition, and numerous studies have compared the olfactory sensitivity of IEI patients and healthy controls (matched for age and sex). The results consistently show no difference between the groups in odor-detection thresholds. In this strict sense, IEI sufferers are no more sensitive to odor than anyone else.

There are some differences in how IEI patients and healthy

people respond to odor, however. For example, patients find the rosy scent of phenylethyl alcohol less pleasant than nonpatients, and they are more likely to report eye/nose/throat irritation in response to it. In another test, IEI patients and controls were confirmed to have similar levels of odor sensitivity. They were then exposed for ten minutes to unscented air or to air with a barely detectable level of 2-propanol (rubbing alcohol). Only 10 percent of normal volunteers reported physical symptoms in either condition. In contrast, 30 percent of the patients reported symptoms to both scented and unscented air. This exaggerated subjective response implies a difference in cognitive processing rather than a change in sensory perception. In other words, a patient's brain intuits harm from a sensory message that causes no alarm in a healthy person.

THE NATURAL HISTORY of odor aversions helps put IEI in perspective. Even the most innocuous scent becomes objectionable if it reminds us of an unpleasant experience. Take the case of Izabella St. James, a former girlfriend of Hugh Hefner who did not enjoy her time at the Playboy Mansion. It was apparently Hef's habit to prepare for festivities in the bedroom by coating himself in baby oil. To this day, says Ms. St. James, the smell of baby oil makes her gag.

Then there is Rolf Bell, a tall, athletic guy in his mid-fifties. When he was six or seven years old, his family visited Mount Lassen in Northern California. They stopped for a picnic at Bumpass Hell, a geothermal area full of boiling mud pots and steaming fumaroles. His mother had prepared egg-salad sandwiches for lunch. After eating his amid clouds of sulfurous steam reminiscent of rotting eggs, little Rolf was left with a permanent olfactory aversion: he hasn't eaten egg salad since.

We sometimes create odor aversions in a misguided attempt to avoid truly bad smells. It's a common impulse to mask the smell of decay with a strong and less objectionable odor. The men who collected the bodies of those killed in the 1900 Galveston hurricane were encouraged to wear bourbon-soaked handkerchiefs over their faces, or to smoke strong cigars. Similar advice was given to personnel in the American Graves Registration Unit who searched the European battlefields of World War II for the remains of U.S. servicemen. Sadly, experience shows that a masking scent may become linked to the emotional trauma of body retrieval duty. Today's military personnel are told not to use cologne to cover the stench.

In January 1987, in the outskirts of Hesperia, a town in San Bernardino County, northeast of Los Angeles, sat a nondescript aluminum building on an asphalt-and-dirt lot surrounded by a chain-link fence topped with barbed wire. The owner of a nearby business noticed flames shooting out of the building's smokestack. What really grabbed his attention was the smell of the smoke: something he hadn't smelled since his U.S. Army unit walked past the ovens at a liberated concentration camp in Germany more than forty years earlier. It was the sickening, strangely sweet odor of burning human flesh. His phone calls to local officials began an investigation that uncovered the largest funeral home scandal ever in Southern California, a grim story of stolen body parts and gold fillings and illegally commingled remains.

These are the smells people can't forget, even if they want to. I'm not talking about clove oil reminding one of a visit to the dentist's office; I'm talking about the extreme edge of human experience. Smells associated with intense trauma leave an indelible impression. Take the case of a fire department paramedic who was called to treat a garage mechanic injured when an automobile tire exploded. The paramedic tried mouth-to-mouth resuscitation, but the victim's face was so badly damaged he had trouble locating the mouth. The victim vomited on him and died. The paramedic was found hours later, sitting in a daze in his car in the middle of an intersection. The smell-linked trauma haunted him for years—whenever he encountered a foul odor, it would bring on a sudden attack of nausea.

The Boston psychiatrist Devon Hinton and his colleagues regularly treat Cambodian refugees, many of whom witnessed atrocities during the Khmer Rouge reign of terror between 1975 and 1979. Olfactory-triggered panic attacks are frequent among these survivors. Innocuous smells such as car exhaust, tobacco smoke, and roasting or frying meat can set off anxiety, dizziness, nausea, and a racing heartbeat. These symptoms are sometimes accompanied by flashbacks to horrific scenes that took place amid the smell of exploding ordnance, and the stench of burned bodies and corpses in open mass graves. Hinton's case summaries vividly record the inhuman savagery that Pol Pot inflicted on his own people, and demonstrate the power of smell to forge a perpetual link to strong emotions.

How to Create an Odor Phobia

Omer Van den Bergh makes people sick. He is a researcher at the University of Leuven in Belgium who has developed a surefire way to induce temporary (yet harmless) physiological distress. He does it by increasing the carbon dioxide level in the air, a simple move with unpleasant consequences. Within twenty seconds of breathing CO_2-enriched air, a person experiences tightness in the chest, a feeling of choking or smothering, a pounding heart, sweating, hot flushes, and anxiety. The symptoms disappear quickly when the CO_2 is reduced to normal levels.

Van den Bergh uses CO_2 to explore the psychological mechanisms of odor aversion. In the basic setup, a volunteer breathes scented CO_2-enchanced air and experiences the usual unpleasant symptoms. When the volunteer returns to the lab the next day and breathes normal air with the same scent, he feels ill again—though there is no physical basis for the reaction. Van den Bergh has conditioned his subjects to feel sick in the presence of an odor, just as

Pavlov conditioned his dogs to salivate at the sound of a bell. Remarkably, all it takes is a single episode of physical distress to turn an odor into a trigger for illness. Van den Bergh calls this process "symptom learning," to reflect the fact that it's a form of associative learning, a basic process by which organisms respond to their environments. Symptom learning works better with malodors, such as ammonia and butyric acid, than with a pleasant, fresh scent like eucalyptus.

Another hallmark of learned aversion is that it spreads from one odor to another in a process known as stimulus generalization. For example, when Van den Bergh conditioned people to become ill at the smell of ammonia, he found they would experience symptoms in a later test when the air was scented with another unpleasant odor, such as butyric acid (smelly feet) or acetic acid (vinegar). The subjects would not become ill to an entirely different smell such as citrus, however. Generalization to related odors can happen as long as a week after the initial event. One consequence of this is that a brief exposure to a sickness-inducing smell leaves a person psychologically vulnerable for days to acquiring additional odor triggers.

If an odor aversion can form after one exposure, and if the aversion can generalize to similar smells, what stops it from becoming a psychological chain reaction? Why isn't everyone gagging all the time? The answer is a phenomenon called extinction. When an illness-associated odor is repeatedly presented without elevated CO_2, the Pavlovian response eventually fades away as the brain unlearns its conditioned response. When the odor no longer triggers symptoms, the response is said to have been extinguished. Therapists use the extinction phenomenon to help people overcome phobias to spiders, closed spaces, and so on; they call it systematic desensitization therapy.

Bad smells are natural candidates for Pavlovian conditioning, but even pleasant ones can trigger symptoms, as we saw in the example of cologne-wearing soldiers on body retrieval duty. Pleasant scents can become triggers under less dramatic circumstances, if they are given the proper psychological "spin." In another of his experiments, Van den Bergh had test subjects read a leaflet beforehand, which discussed chemical pollution and described a patient with MCS. (The text was lifted from an environmentalist website.) In the experiment, the pamphlet's negative spin increased CO_2-induced illness to pleasant as well as unpleasant odors. Thus, even a nice fragrance becomes a trigger for acquired illness if one believes, for example, that its chemical composition is harmful. Van den Bergh sees an irony in this: "warnings and campaigns against environmental pollution, while having important beneficial effects for the environment, may inadvertently facilitate acquiring symptoms to chemicals in the environment and promote the spreading of MCS, mass sociogenic illness, and the like." In other words, we might scare ourselves sick.

Odor-associated symptoms are fertile ground for misinterpretation. If you believe that a particular odor is making you sick, that odor is likely to make you sick even if your original symptoms were caused by something entirely different. Van den Bergh finds that such beliefs better predict odor-induced symptoms than does the person's actual history of odor exposure. People can be made ill by a mistaken belief about a smell. Believing trumps smelling.

IN THE DECADE and a half since the hue and cry in Marin County, many researchers have investigated the MCS/IEI phenomenon, trying to better characterize its symptoms and determine its cause. A review of the large literature on the topic found little evidence that perfume ingredients were the root cause. In fact, it concluded that the toxic-exposure theory of MCS/IEI was dubious: its "hypothesized biological processes and mechanisms are implausible." At the same time, a growing body of scientific evidence points to a nontoxic explanation. Another review found that a psychogenic theory—the idea that the condition originates in the mind as much as the body—is well supported. MCS/IEI may be a psychogenic illness, with patients suffering from the runaway results of symptom learning and stimulus generalization. What's happening to people in the real world may reflect the principles that Pam Dalton and Omer Van den Bergh discovered in the laboratory.

The psychological nature of odor aversions has been known for over a century. "Imagination has, besides, a great deal to do with the supposed noxious effect of perfumes," wrote Eugene Rimmel in *The Book of Perfumes*, in 1871. Rimmel tells of a lady "who *fancied* she could not bear the smell of a rose, and fainted on receiving the visit of a friend who carried one, and yet the fatal flower was only *artificial*." Contemporary research has confirmed the power of the mind. What we believe about a smell, and the malevolent power we attribute to it, alters our sensory perceptions and our physiological responses. This shouldn't come as a surprise: we believe that scent can makes us sexy, relaxed, or alert. This is merely the other side of the coin.

The psychogenic hypothesis doesn't sit well with some IEI patients. They believe their problem is caused by chemicals and nothing else, and they resent any suggestion that some of the problem may be in their heads because it implies their suffering isn't real. The good news for them, if they will only hear it, is that the psychogenic hypothesis points to a treatment and to the hope of a happier life.

From Sacrament to Sacrilege

Fear of fragrance is one of those currents that flows through society like an underground stream. Fed by a mix of well-meaning sympathy, honest confusion, and alarmist hype, it bubbles to the surface here and there, with ironic results.

> *you love righteousness and hate wickedness.*
> *Therefore God, your God, has anointed you*
> *with the oil of gladness above your fellows;*
> *your robes are all fragrant with myrrh*
> *and aloes and cassia.*

—PSALMS 45:7

Church members who are wearing scented products, hair sprays, freshly dry-cleaned clothing, or clothing that was cleaned with fabric softeners, or who have been in a smoky room, will significantly contribute to indoor air pollution.

—*Accessibility Audit for Churches, A United*
Methodist Resource Book about Accessibility

Now, can it be possible that in a handful of centuries the Christian character has fallen away from an imposing heroism that scorned even the stake, the cross, and the axe, to a poor little effeminacy that withers and wilts under an unsavoury smell? We are not prepared to believe so. . . .

—MARK TWAIN, *About Smells* (1870)

I Smell Dead People

HARRY (BILLY CRYSTAL): Suppose nothing happens to you. Suppose you live there your whole life and nothing happens. You never meet anybody, you never become anything, and finally you die one of those New York deaths where nobody notices for two weeks until the smell drifts into the hallway.

SALLY (MEG RYAN): Amanda mentioned you had a dark side.

—*When Harry Met Sally . . .* (1989)

Harry was definitely on to something. The "New York death" is a staple of tabloid journalism. The basic story is always the same: police respond to a neighbor's complaint about a foul odor and discover the body of someone who died alone and unnoticed days or even weeks before. What gives these episodes a typical New York edge is the undertone of alienation and impersonality in a city where people literally live on top of each other—you have to rot before anyone notices your absence. The New York death reveals the city at its worst. In the Bronx in 2004, neighbors heard the sounds of a "battle royal" coming from the apartment of an ex-con who abused drugs, alcohol, and women. Nobody intervened. Nobody called the cops. Two days later the building superintendent phoned police "to report a foul odor." They found the ex-con and a woman dead inside the apartment. Call it the Eau de Kitty Genovese effect.

A New York death can happen anywhere. In Chicago an elderly couple committed suicide with vodka and sleeping pills in a posh Harbor Point Towers apartment. Their bodies were discovered by police only "after residents complained of a foul odor" days later. Near Houston, police found an elderly couple dead in their home "after a neighbor reported a foul odor coming from the house." They had died several weeks earlier, around the time that Adult Protective Services had visited but left when no one answered the door.

The key elements of the New York death are so ingrained in our national consciousness that they have the potential to create embarrassing misunderstandings. After being acquitted in the slayings of his ex-wife, Nicole, and her friend Ronald Goldman, O. J. Simpson moved from Los Angeles to Florida. In 2000 and 2001 he made the news there for various run-ins with the law. He also began dating an attractive young blond woman named Christie Prody. In January 2002, a next-door neighbor noticed "a foul smell" emanating from Prody's apartment and realized she had not seen the woman in about a month. She put two and two together and called the Miami police. They too feared the worst, and had firefighters break into the apartment. Inside, they found no sign of Prody, but they did discover the badly decomposed body of her pet cat. The missing-persons unit was notified and Simpson was questioned. When he reached his girlfriend on the phone, with police present, the matter was resolved. Prody had been out of town for a month and a half, and her cat had starved to death.

One story of stinky corpses has taken on mythic proportions. The "body in the bed" urban legend involves motel guests in Las Vegas who complain of a foul odor in their room, only to discover the next morning that they've slept in a bed with a corpse hidden in or beneath it. Sadly, there is very little mythical about it, other than its being set in Las Vegas. In the past twenty years, odor complaints by motel guests have led to the discovery of murder victims in Atlantic City, New Jersey; Pasadena, California; Alexandria, Virginia; Mineola, New York; and Kansas City, Missouri. Everywhere, it seems, except Sin City.

A common feature of body-in-the-bed incidents is that the tell-tale odor doesn't appear until several days after the murder. A typical case, for example, involved Richard "The Iceman" Kuklinski, the hit man immortalized in an HBO television documentary. According to crime writer Katherine Ramsland, Kuklinski killed one of his victims in a by-the-hour motel on Route 3 near the Lincoln Tunnel in North Bergen, New Jersey. Kuklinski fed the man a cyanide-laced hamburger, and his accomplice strangled him with a lamp cord for good measure. They hid the body under the bed, where it wasn't discovered until the fourth couple to rent the room complained of an odor.

Why does it take motel guests so long to recognize the stench for what it is? Part of the answer lies in biology. In the early 1960s, a Clemson University graduate student named Jerry Payne worked out the detailed chronology of bodily decay that is now the basis of crime-scene forensic investigations. (For example, he pioneered the identification of the stages of insect material—eggs, larvae, and adults—to help determine time of death.) In outlining six stages of postmortem decay, Payne did for death what Elisabeth Kübler-Ross did for dying. Denial, anger, bargaining, depression, and acceptance are followed by fresh, bloat, active decay, advanced decay, dry decay, and remains.

With the exception of the first, each of Payne's stages has a characteristic odor profile. Stage two (bloat) begins on the second day postmortem and lasts one or two days, depending on environmental conditions. Gut bacteria produce sulfur dioxide, and the skunklike smell is often mistaken for a natural-gas leak. (This gives rise to an entire subgenre of New York deaths discovered when a landlord calls utility workers to check out a gas leak. Sometimes there is an ironic twist. In the South Bronx in 2002, an apartment building superintendent and a Con Ed worker sniffing for a gas leak found three people bound and stabbed to death. The telltale gas smell was, in fact, caused by gas. The killers had left the oven open and turned on, and votive candles burning in the living room, hoping that an explosion would obliterate evidence of their crime.)

Active decay—stage three—brings the intense stench of putrefaction. Body tissues liquefy and ferment, giving off a paradoxically sweet smell (and drawing a cheerier crowd of insects such as bees and butterflies). By day six (advanced decay) the breakdown of amino acids produces the accurately named chemicals cadaverine and putrescine, and the superoffensive smell of rot is replaced by an ammonia-like scent. (Cadaverine is sometimes found in bad breath.) Dry decay, which begins about a week postmortem, has a smell reminiscent of "wet fur and old leather." The final stage, like the first, is nearly odorless. All that's left are teeth, bones, and hair.

With this chronology of stench in mind, we can understand why it takes days before motel guests complain. Lynn Nakamura and her brother Dennis Wakabayashi checked into a Travelodge in Pasadena, California, in July 1996. They didn't like the first room they were given, and when the next one proved to have an off-putting odor, they were reluctant to ask for yet another. Two days after they checked out, the motel manager found the body of a murdered young woman hidden in the wooden platform under a twin bed. How could people picky enough to ask for a new room tolerate the smell of death? Easy: they rationalized the problem away. Dennis Wakabayashi said that, to him, the room "smelled like kim chee."

A DECOMPOSING VICTIM is a special challenge for the murderer who lives at the scene of the crime. Most perps fold after a couple of days. From the *New York Daily News*: "A 65-year-old woman was shot dead by her husband and left to rot in the basement of their Staten Island home for two days, police sources said yesterday. The slain woman's 67-year-old husband called cops yesterday because the stench of her decomposing corpse became unbearable." In Foley, Alabama, a thirty-nine-year-old mentally handicapped man died of malnutrition after being kept in a lightless room for ten years by his mother and stepfather. They left his body there for several days, until they could no longer tolerate the smell, at which point they called 911 and were charged with murder.

Some perps are made of tougher stuff. In Tucson, Arizona, a man was found living in an apartment with a the body of a woman who had been dead for almost two years. Police investigated after (yes, you guessed it) "neighbors complained of a foul odor." The man had been paying the rent with the dead woman's checkbook. He told a nosy maintenance man that the odor came from food that had spoiled during a power outage. This guy should be nominated for a Norman Bates Award. Another nominee might be the woman found wandering incoherently inside a Wal-Mart store in Palm Coast, Florida. Sheriff's deputies found her after shoppers complained about a foul odor coming from a car in the parking lot. She and her sixty-five-year-old mother had been on a road trip from Covington, Oklahoma. According to the medical examiner, the mother had died about five days earlier, but the woman had kept driving.

In my opinion, the Edgar Allan Poe Award for the most macabre tale of bodily decay goes to Aron Ralston, the young hiker whose arm got wedged by a boulder as he was climbing a rock face. Stuck out in the wilderness, Ralston could do nothing as his injured limb turned gangrenous, and he had the mind-boggling experience of smelling the rotting of his own flesh. He solved the dilemma by self-amputating the arm, and happily lived to tell the tale.

CHAPTER 7

The Olfactory Imagination

It is never the thing but the version of the thing:
The fragrance of the woman not her self

—Wallace Stevens,
"The Pure Good of Theory"

EARLY IN MY CAREER, I WANTED TO EXPLORE THE psychology of smell, so I decided to do a free-association experiment. The test design was simple: have someone sniff from a squeeze bottle and say the first thing that comes to mind. My main concern in planning the experiment was data reduction—how to record, transcribe, and code the expected torrent of words. I envisioned a panel of judges who would rate the transcripts for emotional content and imagery. I needn't have worried. When given a lemon odor, most people told me, "It smells like lemon." Any particular kind of lemon? "Not really. Just . . . lemon."

So much for the free-association approach. In my naive enthusiasm, I had underestimated the Verbal Barrier. The average person becomes tongue-tied when trying to describe a smell. The reason for this, according to a variety of pundits, is that we have a limited vocabulary for smells. We could describe them better if only we had more words for them. As an explanation, this one is pretty weak. Tar, fish, grapefruit—every smelly thing in the world is a potential adjective. Add to these the names of brands with iconic scents: Play-Doh, Vicks VapoRub, Dubble Bubble, and WD-40. Clearly, there are plenty of words for smell. This means that the Verbal Barrier is not a vocabulary problem, it's a cognitive problem. The words are there, but we have a hard time getting to them.

Psychologists have a name for this slipping of the mental gears: they call it the "tip-of-the-nose" phenomenon. You recognize an odor but can't come up with its name. Tip-of-the-nose happens in real life, but not that often. We are rarely forced to name a random odor with no practice, no context, no prompting, and no multiple-choice options. Yet that is precisely what sensory psychologists ask people to do all the time. Not surprisingly, scores on laboratory tests of stone-cold odor naming are abysmally low. (Researchers give credit for "near misses," such as calling strawberry raspberry, but easy grading doesn't change the overall result.)

Putting a name to a random odor is tough, but the annoying thing about the tip-of-the-nose phenomenon is that you *know* you know the name of the odor. According to the sensory psychologists Harry Lawless and Trygg Engen, the problem is a failure to retrieve verbal information that would lead you to the name. A person trapped in this state of suspended olfaction can name a similar odor about half of the time, but can't think of a word with similar meaning to the odor name. Lawless and Engen could knock loose the correct name in 70 percent of cases by reading the person a definition of the smelly substance. Access to semantic information breaks the tip-of-the-nose spell.

I'm convinced that we make too much of our poor ability to describe smells. The grim reports come from psychology labs, where smells are stripped of context, put in bottles, and given code numbers. Think how hard it would be to verbally describe colors under similar conditions. Interior decorators have fifty-seven words for white, while the rest of us get by with "bright white" and "off-white." Yet for some reason commentators don't moan about our small color vocabulary. In both cases, regular people have enough words and context to get the job done.

The Three Traits of Olfactory Genius

It's a depressing fact that nearly every analysis of putting smells into words—by scientists and pundits alike—stresses weakness and incapacity. The conventional wisdom is oddly anti-intellectual: it seems to deny smell a place in the life of the mind and dismiss its contribution to art and literature.

Yet some writers and artists manage to create works of art in which we recognize our olfactory experience. They invest smells with meaning. They turn an odor into a symbol, a clue to a character's personality, or the atmosphere of a time and place. What do these artists have that the rest of us do not?

I'd like to challenge my academic friends to stop giving random odors to college sophomores in the psychology lab, and start observing odor fluency where it happens naturally—in creative people actively engaged with smell. We need to take a fresh look at how they express olfactory experience in their finished work and at the role of smell in the act of creation. As a first step toward characterizing olfactory genius, we can look for the psychological traits of the olfactively minded artist. I'll kick things off by proposing three of them: awareness, empathy, and imagination.

Let's begin with awareness. Charles Darwin was a great field biologist because he was a careful observer. He was also attuned to smell. Both talents are in evidence in this passage where he describes animal musk: "The rank effluvium of the male goat is well known, and that of certain male deer is wonderfully strong and persistent. On the banks of the Plata I have perceived the whole air tainted with the odour of the male *Cervus campestris,* at the distance of half a mile to leeward of a herd; and a silk handkerchief, in which I carried home a skin, though repeatedly used and washed, retained, when first unfolded, traces of the odour for one year and seven months." For Darwin, smell was a recordable fact like time, place, and species.

Behavioral clues help us identify the odor-aware person. In Portugal years ago, I was eating dinner at a *pousada*, an old castle refitted into an elegant restaurant and hotel. At the next table was a tall, elderly American, his wife, and a Portuguese gentleman. The tall fellow looked familiar; with a bit of eavesdropping I realized it was John Kenneth Galbraith, the economist and diplomat. At the end of the evening Galbraith followed his guests from the dining room. He paused before a large bouquet of red roses near the door, stooped down, and took a long, contemplative sniff. Here was a guy of impressive achievement who actually did stop to smell the roses.

To portray scent in a believable way and have it resonate emotionally, an artist must be alive to smells in the real world. The odor-aware artist is by nature a scent seeker who finds the smells of things, places, and people intrinsically fascinating. He thinks in smells and finds them to be distinct and almost palpable, not wispy and transparent.

To be odor-aware, a person needs only an adequate nose, not a supersensitive one. Emile Zola, the nineteenth-century French writer, is a case in point. His novels were known for their abundant references to smell. Late in life he agreed to be examined by a panel of physicians and psychologists eager to trace creative genius to "organic" factors. Among other things, they did a thorough work-up of his sense of smell. It turned out that Zola's sensitivity was somewhat below average, but not bad for someone in his mid-fifties. Despite his relatively dull nose, his sense of smell was quite refined—he liked to compare and analyze odors, and did so "with a confidence that always astonished his followers." Zola's memory for odors was especially good, and he was able to bring them to mind more vividly than colors or shapes. The investigating panel concluded that Zola's fictional smells were more the result of a supple olfactory imagination than of nose-skills as such.

True odor awareness is probably not very common. We all know people who are indifferent to scent; smells don't grab them emotionally or intellectually. They don't give a damn what their dish detergent smells like, nor do they spend money on perfume or cologne. According to a consumer survey, 23 percent of the population is "apathetic" about perfume and doesn't buy much. At the other end of the spectrum is the 11 percent of the population who are "fragrance fanatics." They own a large wardrobe of scents, which they wear according to season and mood. Let's assume for the moment that artistic talent and olfactory awareness are statistically unrelated. Based on the survey results, we would expect about a quarter of all artists to be indifferent to smell and therefore unlikely to use it in their work. Likewise, only one artist in ten will be a scent-head.

Odor awareness by itself doesn't make one an olfactive genius. Consider the short, messy life of grunge rocker Kurt Cobain. According to the critic Tom Appelo, Cobain's personal journals were riddled with scent images: the lingering *Obsession* of a girlfriend, for example, or Courtney Love's perfume on his pillow. The biographer Charles Cross thinks Cobain was preoccupied with smell. His favorite book was Süskind's *Perfume: The Story of a Murder*, which he read twice. (One wonders whether the hero's suicide fascinated him as much as the smell angle.) However strong his personal fascination with scent, there is little to show for it in his music. The Nirvana anthem *Smells Like Teen Spirit* is an exception. It was inspired by an incident in which friends taunted Cobain about smelling like his girlfriend's deodorant. Kurt Cobain may have been a scent-head, but that didn't make him an olfactive artist.

THE SECOND TRAIT of olfactive creative genius is empathy: a feel for how other people experience smell and respond to it. One might think perfumers are good at this, but it is not necessarily so. The perfumer works in regal isolation. Marketers enter on bended knee with the latest trend forecast, focus-group summary, and consumer test data. A perfumer seldom meets his public. On the other hand, Eric Berghammer revels in his public. He is creating an entirely new artistic medium from scent; he is the world's first Aroma Jockey. This young Dutch artist, who goes by the stage name Odo7, has been "live-scenting" clubs, music venues, and commercial events all over Europe. His tools are simple: braziers and hot-water baths to get the scent into the air, and fans to push it into the audience. In a dance club, Odo7 synchronizes his performance to the DJ's music selections in sets that can last up to two and a half hours. His on-the-job experience makes him an expert in olfactory empathy: from his stage platform he observes how the crowd reacts and he can change the vibe on the dance floor at will. Even in this emotion-laden setting he finds ways to play on smell meaning. He can get laughs from a crowd by wafting baby-powder scent during a heavy-metal tune. Originally a graphic designer and illustrator, Odo7 has shifted paradigms completely. He now translates mood and meaning into scent instead of images. One admires his brass: perfumers would never dare to perform in public.

THE THIRD TRAIT of olfactive genius is a well-developed olfactory imagination. Imagination lets the smell-minded artist translate between the senses and invent new ways for scent to speak to the mind and the emotions.

At the core of olfactory imagination is skill at mental imagery. We can bring to mind an odor the same way we imagine a visual scene. With my colleagues Sarah Kemp and Melissa Crouch, I found a way to measure this ability. We translated a well-validated test of visual mental imagery into olfactory terms. Instead of imagining a specific scene (a lake in the woods, for example) and rating how vivid it appeared in your mind's eye, we asked people to think of an odor (a barbecue, for example) and rate its mental vividness. Compared with civilians, perfumers and other fragrance professionals had more vivid smell imagery, but the same degree of visual imagery. Other researchers have used our test to show that olfactory imagery ability is linked to superior odor perception. It is likely that similar brain areas underlie olfactory imagination and real perception.

AFTER IMAGINING an olfactory effect, the artist has to create it for the public to experience. The stage has always been a favorite experimental playground for the olfactively minded artist. The innovative American director and stage designer David Belasco was an early adopter of olfactory special effects. In 1897 he directed a play set in San Francisco's Chinatown. His staging impressed the *New York Times*: "The senses of sight, hearing, and smell are violently appealed to for the sake of creating an illusion; for the perfume of Chinese punk fills the theatre and the music is as Chinese as possible." The critic for the *New York Journal* didn't buy it: "The entertainment last night began with small whiffs of sickening, nauseating odor that was burned for atmospheric and not for seweristic reasons. . . . The theatre was bathed in this hideous tinkative odor of incense, and during the long overture, you sat there getting fainter and fainter."

Belasco was not discouraged. In 1912 he created a detailed stage replica of a Child's Restaurant (a then-famous New York chain), complete with a working stovetop on which the restaurant's specialty pancakes were prepared during the show. For a melodrama set in a forest of the Canadian Northwest, he strewed pine needles on the stage floor. Aroma was released as the actors crushed them underfoot.

Theatrical scent today rarely ventures beyond Belasco-style

realism. Incense and cooking food are popular effects, but nonliteral atmospheric scents are rare. The campy use of smell, as when Britain's National Opera handed out scratch-and-sniff cards before a performance of *Love for Three Oranges,* leads some directors to avoid odor for fear of wallowing in kitsch. Aroma design remains an intriguing possibility for the theatre; it can be unique or as trite as any other aspect of staging.

The husband-and-wife design team of Charles and Ray Eames created some of the most beautiful (if uncomfortable) pieces of furniture in the twentieth century. Less well known is that they were pioneers of olfactory multimedia. In 1952 they created a show about "communication" for the University of Georgia. *Time* magazine's William Howland called it "one of the most exciting things I have seen, heard and smelled in many years." The show used three slide projectors, two tape recorders, a movie with its own soundtrack, and "a collection of bottled synthetic odors that were to be fed into the auditorium during the show through the air-conditioning ducts." Charles Eames wanted to overstimulate the audience: "We used a lot of sound, sometimes carried to a very high volume so you would actually feel the vibrations. So in the sense that we were introducing sounds, smells, and a different kind of imagery, we were introducing multimedia. We did it because we wanted to heighten awareness." Eames liked the results: "The smells were quite effective. They did two things: they came on cue, and they heightened the illusion. It was quite interesting because in some scenes that didn't have smell cues, but only smell suggestions in the script, a few people felt they had smelled things—for example, the oil in the machinery." Edwin E. Slosson would have been proud; if you cue them with sights and sounds, the audience will create the smells in their own heads.

THE THREE TRAITS of the smell-minded artist find their greatest creative expression in the field of literature. We all know the power of the printed word to conjure images as we read; less well known is that written descriptions evoke appropriately scaled mental images of light, sound, and smell. For example, the phrase "a very very bright light" produces a brighter mental image than the phrase "a weak light." Similarly, a written description allows a reader to accurately imagine a smell's intensity and character. Further, merely reading an odor-related word is enough to activate olfactory regions of the brain. According to an fMRI brain imaging study, "odour words automatically and immediately activate their semantic networks in the [brain's] olfactory cortices." Despite the much-discussed Verbal Barrier, it would seem that olfactory prose offers a potent channel of communication.

Anyone can drop a smell cliché into a story, yet only a few authors bring a true olfactory sensibility to their work. In a letter to *The Nation* in 1914, the English professor Helen McAfee mourned the fact that the smells in contemporary American fiction were all clichés: "For example: the complementary smell of a New England spinster story, lavender; of a tale of camp life, pines; of a June romance, roses." She praised Russian authors like Chekhov and Dostoyevsky, whose smells "are keen and fresh . . . not dragged in simply for form's sake." When smell is used in this way, she wrote, "the impression on the reader is correspondingly deep." Inspired by Professor McAfee, let's take an unapologetically nasocentric point of view and ask, Who are the writers that bring an olfactory dimension to their work, and how do they make it succeed?

"This all started on a Saturday morning in May, one of those warm spring days that smell like clean linen." So begins Anne Tyler's novel *Ladder of Years*, about a woman who walks away from her family during a beach vacation in Maryland and starts a new life of anonymous domesticity. Tyler plays on the theme of interchangeability—of people, places, and entire lives—and supports it with deliberately generic odors. A doctor's office smells like a "mixture of floor wax and isopropyl alcohol," a town library exudes "a smell of aged paper and glue," and so on. The heroine notes these familiar odors, but they don't touch her emotionally.

The smell of freshly baked bread drifts through Jay McInerney's *Bright Lights, Big City* amid constant references to snorting coke and its nasal complications. The story begins with the protagonist's reminiscence of waking up to the smell of an Italian bakery in Manhattan's West Village, in an apartment he shared with an old girlfriend, now his ex-wife. The scent turns up in a passage about his mother at home, and again when his sympathetic coworker Megan buys him a loaf to see him through a coke-fueled downward spiral. At the burned-out end of a nonstop weekend of partying, he trades his Ray-Bans to a bakery deliveryman who tosses him a bag of hard rolls. The aroma returns in the book's famous last lines: "You get down on your knees and tear open the bag. The smell of warm dough envelops you. The first bite sticks in your throat and you almost gag. You will have to go slowly. You will have to learn everything all over again." (Skeptical readers might object: Wouldn't heavy use of Bolivian marching powder have devastated the hero's sense of smell? After all, long-term snorting results in sniffling, nasal crusting, ulceration, bleeding, postnasal drainage, and, most spectacularly, a perforated septum. The only study of smell in cocaine abusers found that ten of eleven had a normal sense of smell; even the patient with a perforated septum could smell just fine.)

ONE AMERICAN WRITER, Nathaniel Hawthorne, embodied all three traits of olfactive genius. His novel *The House of the Seven Gables* is filled with smells. Here is the New England village feast celebrating the completion of the house: "The chimney of the new house, in short, belching forth its kitchen-smoke, impregnated the whole air with the scent of meats, fowls, and fishes, spicily concocted with odoriferous herbs, and onions in abundance. The mere smell of such festivity, making its way into everybody's nostrils, was at once an invitation and an appetite." Clearly, Hawthorne was a man who liked to eat.

In *The Scarlet Letter,* Hawthorne describes the Inspector of the Custom-House, a man who was the son of a Revolutionary War colonel. The Inspector was remarkable for "his ability to recollect the good dinners which it had made no small portion of the happiness of his life to eat." Not only could he recall the sensory details; he could vividly summon them up for the appreciation of others: "His reminiscences of good cheer, however ancient the date of the actual banquet, seemed to bring the savor of pig or turkey under one's very nostrils." We see the Inspector before our eyes and work up an appetite just reading about him.

"Rappaccini's Daughter" is perhaps the best smell-based story in American letters. Set in Padua, Italy, around the turn of the seventeenth century, Hawthorne's tale concerns a medical student who becomes infatuated with the beautiful daughter of Dr. Rappaccini. The dour physician breeds poisonous plants, and has deliberately raised his daughter in close contact with them so that she is not only immune to their effects, but has become a repository of their toxins. She exudes an intoxicating and toxic fragrance. As the student courts her, he too becomes saturated with the debilitating scent. The story ends tragically when a rival doctor provides the lovers with an antidote.

Hawthorne was keenly aware of smells, he had an empathic sense of how they affected others, and he could express them in a sustained way in the course of wonderful stories. Although he was descended from austere New England Puritans who rejected sensuality, Hawthorne himself was blessed with a joyful nose.

The Creative Spark

There is a well-worn anecdote about smell and literary creativity. It's about the German poet and playwright Friedrich Schiller. One day his good friend Goethe paid him a visit. Goethe was cooling his heels in Schiller's study when he noticed an overpowering and somewhat nauseating odor. He asked Frau Schiller about it, whereupon she pulled open a desk drawer filled with rotten apples. She told Goethe that her husband couldn't get the creative juices flowing without a whiff from the old apple stash. Whether she rolled her eyes when she said this is not recorded.

This story is supposed to illuminate the psychology of olfactory inspiration, but that's always seemed a bit of a stretch to me. Did Schiller write particularly well or often about apples? Did he have a theory linking apple scent and inspiration? Did he ever try peaches? As far as I can tell, Schiller's apple-sniffing was nothing more than a compulsive warm-up ritual.

There are better places to seek the link between scent and creativity. A good place to start is with the American poet Emily Dickinson (1830–1886). This near-recluse lived her entire life at the family home in Amherst, Massachusetts. She was knowledgeable about botany and obsessed with flowers, of which she grew many kinds on the property and in an indoor conservatory. Cultivating flowers was a hobby for many women of her time, but unlike them Dickinson could not have cared less about showy, scentless orchids. Her exclusive passion was scented flowers. Her favorites make an impressive list: French marigold, mignonette, peony, primrose, Sweet Sultan, Sweet William, roses of various kinds, lilac, mock orange, honeysuckle, jasmine, heliotrope, and sweet alyssum. Dickinson was not into subtlety; she preferred the strong perfume of tropical jasmine and ripe "Bourbon" roses. Her conservatory was saturated in scent. Given the Victorian sensibilities of the time, these lush blossoms were considered too suggestive for the drawing room. Instead she placed pots of them in her bedroom and next to her writing desk.

Not surprisingly, flowers are a major theme in her work; one in five of her poems refers to flowers in some way. She was known around town for sending people her odd little poems tied to a home-grown bouquet. This is how most of her poems became public during her lifetime; very few were published. Once her complete works were issued in 1955, Dickinson finally was showered with critical praise, especially for the way her poems displayed a "cultivation of emotional intensity."

Camille Paglia challenged this admiring consensus in 1990, when she portrayed the poet as a death-obsessed vampire feeding on the emotional intensity of others. Calling Dickinson "the female Sade," Paglia pointed to the poet's "unrecognized appetite for murder and mayhem," and described her poems as "screenplays of agony and ecstasy where someone is tortured, dying, transfigured." This reassessment was so ferocious that I didn't believe it at first—but then I browsed through Dickinson's poems. In addition to flowers, her poetry is mostly about bees and death. Of 1,175 poems, roughly 400 are about flowers. Yet only two mention fragrance directly ("spicy carnations" and an "aromatic" pink); a handful of others allude to it. This is weird: her life revolved around the growing of scented flowers; she wrote her poetry surrounded by them; she even used them as a metaphor for creativity—so why doesn't she describe their scent in verse?

The answer is that Emily Dickinson didn't inhale fragrance like a normal person—she drank it. In her poems, the scent of flowers is nourishment. Describing the scent of spring, she calls herself "a drinker of Delight." She gets drunk on fragrance: "Inebriate of Air—am I— / and Debauchee of Dew." She and the bee "live by the quaffing," she on Burgundy, the bee on clover nectar. She raises flowers in order to consume their fragrance, which fuels her creative powers. There's no denying it: Emily Dickinson was a fragrance vampire.

In Amherst one day, Miss Dickinson cut some bee balm and put a pot of jasmine out in the rain. Purely innocent actions had anybody noticed them. But inside the S&M hothouse of her imagination, these become: "Kill your Balm—and its Odors bless you / Bare your Jessamine—to the storm / And she will fling her maddest perfume / Haply—your Summer night to Charm." In other words, death and violent exposure lead to blessings and nocturnal ecstasy. Dickinson's flowers yield up their scent in the act of dying: "And even when it dies—to pass / In Odors so divine / Like Lowly spices, lain to sleep / Or Spikenards, perishing." Dickinson sucks the scent-soul out of a dying blossom and begins scribbling lines of verse. "They have a little Odor . . . spiciest at fading." To the ghoulish Belle of Amherst, the fragrance extracted at the moment of death was the tastiest.

I now think Camille Paglia got it right: our poet had an appetite for murder and mayhem. "Essential Oils—are wrung / The Attar from the Rose / Be not expressed by Suns—alone— / It is the gift of screws." Ouch! Dickinson tortured the perfume out of flowers.

This casts a sinister new light on the poet's album of pressed flowers, lovingly preserved in the Emily Dickinson Room of the Houghton Rare Book Library at Harvard. Scholars celebrate it as a beautiful record of her passion for flowers. I think the album is a creepy thing—it houses the trophies of a serial killer.

THE COMPOSER Richard Wagner was another fragrance freak. He used mass quantities of scent in his daily bath and dusted his outrageous silken and fur outfits with aromatic powders. His personal letters are filled with discussions of perfume. The scholar Marc Weiner points out that Wagner's "fetishistic fascination with odor" carried over into his operas. When the word *Duft* (fragrance) appears in a libretto, the context is almost always titillating, dangerous, and erotic. Beautiful *Duft* scents the air at every disguised suggestion of sibling incest, as in the first encounter between Sieglinde and Siegmund in *Die Walküre*. Noticeably *Duft*-less are socially acceptable unions such as the bourgeois marriage of Eva and Walther von Stolzing in *Die Meistersinger*. To lesser beings such as the dwarves

Alberich and Mime in the Ring cycle, or the cobbler Beckmesser in *Die Meistersinger*, Wagner assigns unpleasant smells. (Beckmesser stinks of the pitch he uses as shoe-black, which tags him with a satanic theme.) In *Siegfried*, a trilled theme on the piccolo serves Mime (in Weiner's refined phrase) as "a leitmotif for abdominal wind."

Me Smell Sexy

The night is cool. I feel a slight chill. The atmosphere is heavy with the odor of flowers and of the forest. It intoxicates.

—LEOPOLD VON SACHER-MASOCH, *Venus in Furs*

Regular smells can become eroticized. The association of fragrance and forbidden sex was a robust literary theme in the nineteenth century. Witness Leopold von Sacher-Masoch (1835–1895) and his 1870 novel about the whip-wielding mistress Wanda, which has immortalized his name: masochism. Early on, the narrator Severin tells of his fascination with Wanda and his growing fantasy of submission to her. His erotic intoxication with Wanda is drenched in scent. A typical observation: "A sultry morning, the atmosphere is dead, heavily laden with odors, yet stimulating."

Once Severin agrees to be Wanda's slave, everything changes. They travel to Florence; in the carriage ride to the train station she is playful, but her warmth and scent are already receding: "she even gave me a kiss, and her cold lips had the fresh frosty fragrance of a young autumnal rose, which blossoms alone amid bare stalks and yellow leaves and upon whose calyx the first frost has hung tiny diamonds of ice."

As the domineering Wanda becomes more remote, odors become coarse and repellant to Severin. Wanda rides in a first-class train car, but makes Severin sit with the plebes: "Then she nodded to me, and dismissed me. I slowly ascended a third-class carriage, which was filled with abominable tobacco-smoke that seemed like the fogs of Acheron at the entrance to Hades." Here he has "to breathe the same oniony air with Polish peasants, Jewish peddlers, and common soldiers." After a layover in Vienna they proceed to Florence. "Instead of linen-garbed Mazovians and greasy-haired Jews, my companions now are curly-haired Contadini, a magnificent sergeant of the first Italian Grenadiers, and a poor German painter. The tobacco smoke no longer smells of onions, but of salami and cheese." The pungent odors of everyday life crowd out the heady scents of Severin's submissive fantasy. By the end of the story he stands before his original image of the ideal mistress: the statue of the Venus of Medici. In his despair, he sees on the statue "fragrant curls which seemed to conceal tiny horns on each side of the forehead." He has given his soul to a she-devil with a heart of stone. Sacher-Masoch created an olfactory accompaniment for the arc of Severin's story; a descent from heady fantasy into submission and then despair.

WHEN AUTHORS EROTIZE scent, they may reveal something of themselves. Take the American novelist Willa Cather, for example. She never married and lived for long periods with woman friends. Her sexual identity remains ambiguous and is the subject of much speculation in Queer Studies departments. *O Pioneers!* is her 1913 novel about illicit love and a doomed affair on the Nebraska frontier. Its unsentimental heroine, Alexandra Bergson, treats men as fellow workers, never marries, and never consummates a love affair. The many indoor smells of *O Pioneers!*—spirits, pipe smoke, damp woolens, kerosene, and noxious Mexican cigarettes—are all unpleasant, manufactured, and male. In contrast, the outdoor smells are positively emotional and almost erotic. There is the "strong, clean smell" of brown earth in the springtime that "yields itself eagerly to the plow," the spicy odor of wild roses after a rain, ripe fields of corn and wheat, sweet clover, and evening air "heavy with the smell of wild cotton," the "more powerful perfume of midsummer." Alexandra has one romantic fantasy in which she is carried away by a strong, anonymous man: "She never saw him, but, with eyes closed, she could feel that he was yellow like the sunlight, and there was the smell of ripe cornfields about him." Cather's human eroticism (such as it is) is much like her eroticizing of nature. The Queer Studies folks might be missing the point. The nose clues suggest that Cather's sexual orientation was far too diffuse to be captured by either end of the male-female continuum.

WILLIAM FAULKNER was an old man when a student asked him about the many references to scent in his writing. Faulkner replied that "maybe smell is one of my sharper senses, maybe it's sharper than sight." I think Faulkner was saying what he thought the kid wanted to hear. I'm doubtful that smell was Faulkner's sharper sense, because it doesn't add up with anything else we know about him. He was a dapper dresser who apparently didn't wear cologne. There are perfunctory references to lilacs in his early romantic poems. There is not much to suggest that he had a heightened awareness of odor. Nor did he write about scent naturalistically. He used smells a lot, but in a brilliantly contrived way. Faulkner has been called "the most radical innovator in the annals of American fiction." He didn't get this reputation from the precise observations of his "sharper sense"; he got it from a highly original use of smell as metaphor.

Faulkner set his stories in the South, yet he took the stereotypically sweet and romantic scents of wisteria and honeysuckle and turned them into symbols of sorrow and "the inherent tragedy of southern history." He pushed the envelope further in *The Unvanquished*, a novel about young Bayard Sartorius, whose father was a colonel in the Confederate cavalry. At first Faulkner pairs smell with emotion in conventional ways: gunpowder with conflict, and dead roses with a murdered grandmother. It's not until the final chapter—"An Odor of Verbena"—that Faulkner really uncorks the olfactory symbolism. In real life, true verbena (*Verbena officinalis*) is nearly scentless. By talking about it as if it had an aroma, Faulkner forces the reader to see the scent as a symbol of courage and violence.

Faulkner gives the odor of verbena a different strength in each scene. Bayard perceives the "now fierce odor of the verbena sprig" on his jacket as he walks to confront his father's killer. The Southern code of honor demands that he avenge the murder. When the man fires twice, deliberately missing the unarmed Bayard, honor has been satisfied without bloodshed. Bayard returns home and is able to smell the flowers at his father's wake above the now-diminished odor of verbena. We understand that violence is no longer needed; the call for courage has been met. Smells wax and wane in real life; Faulkner's genius was to synchronize the sensory with the symbolic.

His most extended use of smell was in *The Sound and the Fury*, a novel about the breakdown of the Compson family of Mississippi. Faulkner tells it in time-fractured sections and by taking the point of view of different characters, each with their own smells. The mentally defective Benjy perceives the world as a confusing, multisensory jumble; he finds calm in the bodily scents of his caretakers, especially his sister Caddy. Benjy's constant refrain is that Caddy "smells like trees." Their brother Quentin's obsessive, guilt-ridden, and erotically tinged thoughts about Caddy are paired with "the twilight-coloured smell of honeysuckle." When Quentin prepares to commit suicide, the tone of the story changes and honeysuckle is replaced by the harsh smell of gasoline. Jason is the hard and cynical Compson brother who lacks feelings. The stink of gasoline and camphor are the only smells to appear in his story. In the novel's final section, an all-knowing voice completes the story against a depersonalized and oppressive aromatic backdrop: "obscurity odorous of dank earth and mould and rubber," a "faint smell of cheap cosmetics," a "forlorn scent of pear blossoms," and a "a pervading reek of camphor."

Faulkner tried to convince an impressionable undergraduate that this all proceeded from a sharp sense of smell, that he had "no deliberate intent" to make a big deal of smell in his work. But I detect a whiff of bullshit. Masterfully gauged metaphors don't happen by themselves.

A Night at the Opera

Early in 1993, I received a letter from Roland Tec, director of the New Opera Theatre Ensemble of Boston. Tec was producing a new work called *Blind Trust,* a boy-meets-blind-girl story with an improvised score and script. The production was to take place entirely in the dark, with scenes to be accompanied by scent to give a sense of place. Could I help them do this?

I convinced my boss at Givaudan-Roure Fragrances that this was an interesting creative challenge, one that would shower the company with free publicity and position us as a patron of the arts. *Blind Trust* became an official project, and we started designing atmospheres for a pizza parlor, a flower shop, a laundry, and a movie theater. Some of the fragrance development was easy—the flower shop required only a basic floral bouquet formulation with an exaggerated "green" note to suggest stems and leaves. We already had an excellent freshly-pressed-linen accord for the laundry. Pizza and buttered popcorn required extra effort—I crossed corporate boundaries and called the flavor division for help.

With initial fragrance formulations in hand, the next step was to adjust them so they smelled right in a big air space. This is not a concern for fragrance worn on skin, but it's a critical step in developing an air-freshener scent. An oil that smells good on a piece of blotter paper takes on an entirely different character when it fills a room via aerosol or scented candle. The fragrance may "fall apart": one component overwhelms the others, or is lost entirely. To get a sense of how a fragrance will smell in actual use, we test them in small rooms or, in our case, stainless-steel booths.

Within a week or so I was conducting informal scent-booth evaluations of the *Blind Trust* fragrances. Our staff, usually called on to rate the next "Country Meadow" air freshener, were amused to be judging pizza aroma. Still, their comments were useful ("more garlic," "less basil," "find a better cheese note"). When we tested the buttered-popcorn smell one afternoon, people wandered in from all over the building, asking who had microwaved the popcorn.

Blind Trust premiered in the planetarium of the Boston Science Museum on June 5, 1993. Tec's artistic conception demanded that the audience experience everything as a blind person would—by ear or nose only. Instead of dimming the house lights, Tec plunged the room into complete blackness. Instead of a graceful word of welcome, he read aloud the program notes in their entirety. The music began and the singers stood next to the star-projector in the center of the room. Tec's four odor-wranglers stealthily took up positions by the hall's air inlets, located on the walls at head level. Armed with aerosol cans, they waited for their cues to start spraying. It was soon clear that even four cans at once were no match for the planetarium. Odors that were powerful in a living room seemed delicate in a hall this big. Also, the cues weren't always well timed. Too often the smell arrived before the scene had been established, leaving the audience sniffing in puzzlement. Instead of building a multisensory realism, the scent effects sowed confusion. Smelling my contributions in action, I thought we could have improved upon them here and there: the pizza was overly garlicky, and the fresh-linen smell in the dry cleaner's scene was too weak.

In the pitch-black hall, it was hard to know when a scene was over, leaving the audience uncertain when to applaud. At the end of this long and frustrating performance, Tec read the show's entire production credits, thereby destroying whatever sympathy his beleagured audience had left.

The *Boston Globe*'s review was merciless: "Blind Trust: Hold Your Nose." While noting that Tec's troupe had "built a modest reputation for creating new, quasi-improvised operas on themes of political correctness, Cambridge-style," the paper ripped the new production to shreds. The music was "worthless" when not "derivative and mechanical" in the Phillip Glass mode. The improvised singing consisted of "verbal, vocal and harmonic cliché." And the odors, alas, were "confusing and unpleasant."

In the end, Givaudan-Roure did not get the positive press it had hoped for. We didn't even get credit for trying. When a show stinks up a storm all by itself, I'm not sure even the best of stage scents can salvage it. Roland Tec went on write a play and direct a movie. The last time I checked, however, *Blind Trust* was not part of his online biography.

CHAPTER 8

Hollywood Psychophysics

[T]he producers of this film believe that today's audiences are mature enough to accept the fact that some things in life just plain stink.

—from the prologue to *Polyester*

I SAW JOHN WATERS'S FILM *POLYESTER* ON ITS FIRST RELEASE in 1981, in a packed theater in Philadelphia. Like everyone else, I scratched and sniffed my Odorama card as an onscreen character named Francine Fishpaw (played by the obese and outrageous Divine) let one loose under the bedcovers. The audience groaned; we knew what was coming, yet we all inhaled. To this day, Waters delights in his cinematic coup: he tells me "audiences worldwide paid me money to smell a fart."

The idea of smelling a movie has been a joke for so long, it's easy to forget that scented films once played at major venues in New York, Chicago, and Los Angeles. History has not been kind to Smell-O-Vision or its rival, AromaRama; they have been relegated to books like *Arts & Entertainment Fads*, and *Oops: 20 Life Lessons from the Fiascoes That Shaped America*. The *Times* of London wrinkled its editorial nose and called them "cinematic stinkers" and "historic blunders." Smell-O-Vision made *Time's* list of the 100 Worst Ideas of the Century, along with Hair Club for Men, leisure suits, and New Coke. Michael and Harry Medved nominated it for a Golden Turkey Award in the category of "Most Inane and Unwelcome 'Technical Advance' in Hollywood History."

The loud mockery of the pundits strikes me as a cheap shot for a couple of reasons. First, I feel a warm emotional connection to the smelly moments of Hollywood history, perhaps because of my personal role in the odorific failure of *Blind Trust*, or my involvement during the dot-com boom with a startup called DigiScents, Inc., that aimed to bring smell to the Internet via a PC-linked scent generator. Why is it so difficult for critics to believe that people have a sincere interest in the possibilities of scented entertainment? My second reason is a lingering suspicion that the magazines and media professors are missing something important: If it's really such a bad idea, why does the public remain so fascinated by it? I decided to take a closer look for myself, and began spooling through miles of microfilm and talking to people who had experienced Smell-O-Vision and AromaRama for themselves. My goal? To find out whether there was something more to the story than all the snark would suggest.

THE FIRST ATTEMPT to odorize movies dates back to the earliest days of silent film and was the brainchild of Samuel "Roxy" Rothafel (1882–1936), the legendary cinematic impresario who ran New York venues such as the Rialto and the Strand. The lavish movie palace he created and named after himself—the Roxy—became a generic name for cinemas across America. The man helped make Hollywood what it is today, but the story of Rothafel's smelly movie has a few holes in the plot.

According to *Film Daily*, Rothafel "tried the rose bit back in 1906, in a silent-film house he ran in Forest City, Pennsylvania. For newsreel clips of the Pasadena Rose Bowl Game, he dipped absorbent cotton in a rose essence and put it in front of an electric fan." This charming story is repeated in book after book on the history of movies. There's only one problem with it: there was no Rose Bowl game in 1906. The first one was played in 1902; it was such a blowout (Stanford conceded in the third quarter, trailing Michigan 49–0) that the Tournament of Roses gave up on football and ran chariot races for a few years. Football didn't return until 1916 (Washington State 14, Brown 0). So at what movie was Roxy blowing rose essence in 1906? Pasadena had hosted a New Year's Day Rose Parade since 1890, and the Vitascope Company filmed it for the first time in 1900. It's more likely that Roxy scented a newsreel of flower-trimmed floats in the 1906 Rose Parade.

Roxy never repeated his improvised stunt, but it was imitated by others. In 1929 the manager of Boston's Fenway Theatre poured a pint of lilac perfume into the ventilation system; he timed it to hit the audience just as the movie's title—*Lilac Time*—flashed on screen. The same year, an orange scent was dispensed at Grauman's Chinese Theatre in Los Angeles during showings of MGM's *Hollywood Review*; the smell came during a big musical number called "Orange Blossom Time."

SCENTED ENTERTAINMENT as an art form needs something more than a projectionist with a screwdriver and a flask of perfume. At around this time other people were giving serious thought to the artistic and dramatic potential of smell. Aldous Huxley offered a whiff of the possibilities in his 1931 novel *Brave New World*:

> The scent organ was playing a delightfully refreshing Herbal Capriccio—rippling arpeggios of thyme and lavender, of rosemary, basil, myrtle, tarragon; a series of daring modulations through the spice keys into ambergris; and a slow return through sandalwood, camphor, cedar and newmown hay (with occasional subtle touches of discord—a whiff of kidney pudding, the

faintest suspicion of pig's dung) back to the simple aromatics with which the piece began. The final blast of thyme died away; there was a round of applause; the lights went up.

It's a great fantasy: smells arrive at the nose in precisely timed pulses and disappear just as quickly. But as I learned in *Blind Trust*, moving scent through a big space is an inexact art form. Fan-blown air masses move slowly and linger too long; it's easy to end up with olfactory sludge.

There is another problem. Even if a scent organ delivered odors with the brisk precision that Huxley imagined, the audience would have trouble keeping up. Fragrance arpeggios would blow by too quickly for the human nose to perceive distinct notes. (A mouse, on the other hand, might get it. Mice generate a fresh impression of the smellscape with each sniff, and since they sniff several times a second, they can easily keep up.) The human nose works on a longer time scale; it can't follow a smellody the way the ear follows a tune. Anything faster than *largo ma non tropo* would leave an audience in the dust.

Bill Buford encountered a typically sedate olfactory tempo when he worked as a line chef in the kitchen of an Italian restaurant:

> By midmorning, when many things had been prepared, they were cooked in quick succession, and the smells came, one after the other, waves of smell, like sounds in music. There was the smell of meat, and the kitchen was overwhelmed by the rich, sticky smell of wintry lamb. And then, in minutes, it would be chocolate melting in a metal bowl. Then a disturbing non-sequitur like tripe (a curious disjunction, having chocolate in your nose followed quickly by stewing cow innards). Then something ripe and fishy—octopus in a hot tub—followed by overex-tracted pineapple. And so they came, one after the other.

Another obstacle to olfactory cinema is clearing the air between performances. The movie-industry veteran Arthur Mayer found this out in 1933 when he installed the first true in-theater smell system. He had just taken over Paramount's Rialto Theater on Broadway, when he was approached by an inventor who claimed he could deliver scent to an audience in synchrony with a movie. His demo film about a pair of young lovers was accompanied by all sorts of smells. There was a hitch, however, as Mayer recalled:

> The blowers which wafted these odors out with such precision were supposed to waft them back with equal efficiency, but unfortunately this part of the invention had not yet been entirely perfected. The auditorium was so full of a mingling of honeysuckle, bacon and Lysol that it took over an hour to clear the air and for several days afterward there was such a strong smell of those mature apples around that a friend asked me if I was making applejack on the side. It was a long time before I finally lost confidence in the smellies, but my man and I—I had become a zealous partisan if not a partner—could never seem to master the backwards waft.

Mayer didn't name his olfactory accomplice, but a cartoon in his book provides a clue. It shows Mayer in a projection booth, peering down into the house. Next to the film projector is a large device with tubes labeled "rose," "honeysuckle," "Lysol," "ripe apple," etc. The scent tubes lead into ventilating ducts that open into the theater. This arrangement is precisely the system described by John H. Leavell in a U.S. patent issued three years before Mayer met his unnamed inventor. If it was indeed Leavell who installed scent at the Rialto, then despite his short-lived partnership with Mayer, he deserves to be recognized as a pioneer of scented cinema.

In any case, the idea of odorized movies had taken on a life of its own. Walt Disney got excited about it when he was planning *Fantasia* in 1938. He considered floral perfumes for the *Nutcracker Suite*, incense for the *Ave Maria* and *Credo*, and gunpowder to stoke the devilishness of the *Sorcerer's Apprentice* sequence (his conductor,

Leopold Stokowski, was especially keen on this). Disney, while reluctant to give up on such a "great publicity angle," eventually decided to steer away for cost reasons. A 1944 Warner Bros. cartoon called "The Old Grey Hare" followed Bugs Bunny and Elmer Fudd into the distant future; an elderly Fudd reads a newspaper headline in the year 2000: "Smellovision Replaces Television." The Soviet Union, sensing another Cold War technology challenge from the Americans, tried to get in on the act. The Russian movie director Grigory Alexandrov claimed in 1949 that the Soviet film industry "was on the verge of producing smellies," but there is no record they ever did.

The Path to Smell-O-Vision

Smell-O-Vision was the lifelong quest of an obscure Swiss-American entrepreneur and fragrance enthusiast named Hans E. Laube. The saga began in 1939 when Laube, a tall, bespectacled, thirty-nine-year-old advertising executive from Zürich with a flair for invention and a passion for fragrance, developed a theatrical scent system that released multiple smells during a film. Along with financier Robert Barth and movie producer Conrad A. Schlaepfer, he formed a company called Odorated Talking Pictures. As a show-case for their new technology, the partners spent 30,000 Swiss francs (about $101,000 in today's terms) to make an English-language fea-ture film called *My Dream*. Its rudimentary plot included twenty smells: "A young man meets a pretty woman in a park. She disap-pears, but lets fall a handkerchief which diffuses a perfume. On the basis of this smell the man takes up pursuit. The public can also smell along: Rose scent, hospital atmosphere, car exhaust, and finally incense during the wedding of the pair in a Gothic chapel."

The OTP partners unveiled their system at a press conference in Bern on December 2, 1939, garnering a mention in the *New York*

Times in February 1940. Even better, they arranged to have *My Dream* shown in the Swiss Pavilion at the New York World's Fair.

On the evening of Saturday, October 19, 1940, Laube's scented film was shown in public for the first—and evidently last—time in the United States. The film historian Hervé Dumont describes what happened: "At the conclusion of the performance the O.T.P. equipment, along with the only copy of the film, is seized by the American police under the pretext that a similar, patented system already exists in the USA. The promoters stay in town and press various lawsuits in order to get back their material. In vain: Barth dies there, after he—like Schlaepfer—lost his entire investment."

Despite this disaster, Laube refused to quit. He stayed in America during World War II to promote his inventions. Laube pitched supermarket ad displays with smells to accompany slides of food. He developed a device he claimed could release odors in synchrony with a television broadcast—more than 2,000 odors-on-demand available in your living room. Film and television deals continued to elude him, however. He became disillusioned and returned to Europe in 1946.

Enter Michael Todd

Laube, a quiet and intense inventor, might not have gotten very far had he not met Michael Todd, a Broadway impresario and flamboyant force of nature. A risk-taker and a feisty competitor, Todd spent freely on special effects to draw big crowds to his shows, and every one of his hit musicals featured an extravagant set or stage effect.

Yet Todd had more than a showman's interest in special effects: he helped invent and commercialize several movie-making technologies. Todd's Broadway hit, *The Hot Mikado*, was playing at the 1939 World's Fair, and while keeping an eye on the show he met Fred Waller, who was demonstrating an eleven-projector wraparound movie film system called Vitarama. Cinerama, a three-camera, wide-screen format that was projected onto a specially shaped screen, was another Waller invention, and Todd became an investor in it. Expense and complicated technology were no barrier for Todd; his enthusiasm and salesmanship persuaded movie distributors to pony up and install the new equipment. He made a splash with *This Is Cinerama* (1952). Audiences thrilled to a sequence filmed on a roller coaster at Coney Island. It was the IMAX of its time, and eventually led to today's Panavision system.

Mike Todd may have noticed another promising technology at the World's Fair: Hans Laube's Odorated Talking Pictures. It's not clear whether Todd and Laube actually met there, but somehow Todd caught the scent bug. By 1954 Laube was back in America, trying to bring aroma to movies and television. That year he gave a demonstration to Todd and the producer decided to invest in the new system.

In his 1954 application for a U.S. patent, Laube described a device in which odor canisters were placed on a turntable. An electronic scent-track on the motion picture film triggered the turntable, which rotated the desired canister beneath a pickup nozzle, which sucked up scent and pumped it into the theater through tubes attached to the seatbacks. The liquid fragrances were filtered to remove the heavier notes and prevent the scents from lingering too long. To help clear the air between smells, one canister contained an "odor neutralizer." The odors could be played in a fixed sequence or the scent-track could advance the turntable to any desired canister. Laube's idea was that theaters would receive a standard set of odors; if a movie had unusual scent effects, a custom set would be shipped along with the film reels.

By 1955 Laube's career was gaining momentum. He gave a private demonstration of his system at the Cinerama-Warner Theatre in New York, using a short version of *My Dream*. It must have been a success, because he persuaded the Stanley Warner Corporation, which owned the rights to Cinerama, to fund further development. To secure international rights to his invention, he filed a European patent application, and then applied for a second U.S. patent. In May, Laube married for the second time after a month-long romance. In July he received shares in a newly formed company called Scentovision, Inc.

In September 1956, Scentovision held another private demo for industry executives at Mike Todd's Warner Cinema in New York. The 16 mm film ran for eight and a half minutes and used seventeen aromas. *Motion Picture Daily* hinted that Laube's system would be installed in a top theater within nine months, and that Scentovision was negotiating with film producers who wanted to use the process. In November 1957, Laube and a partner were issued U.S. Patent 2,813,452, "Motion pictures with synchronized odor emission," and were mentioned in the *New York Times*.

MICHAEL TODD's first movie, the 1956 blockbuster *Around the World in 80 Days,* capitalized on his marketing strategy of heavily hyped limited openings, and heavily marketed accessory items (the movie's soundtrack album was the first nonmusical soundtrack to earn big money). Early in 1957 he married Elizabeth Taylor—the third marriage for each of them—and a month later the newlyweds attended the Academy Awards, where *80 Days* won the Oscar for Best Picture. With movie profits rolling in, Todd was looking about for his next project, and he felt the time was right for a push into smellies.

Things were finally looking good for Scentovision. Hans Laube had a patent, a prototype system, and a company to promote it. Mike Todd had committed to funding the technology and was considering it for a major movie. Then, on March 21, 1958, Todd was killed when his private plane went down in a storm over Grants, New Mexico.

After the funeral, twenty-eight-year-old Mike Todd Jr. took the reins of his father's production company, where he had been working for years. Though the son had little of his father's charisma and outsized appetites, he was a smart and sociable young man with ambitions of his own. Perhaps hoping to establish himself with a blockbuster new film process, Mike junior threw himself and his company's resources behind a smell movie project called *Scent of Danger*. He signed Hans Laube to an exclusive, long-term contract and lent him the company's New York warehouse space to work in, and the Cinestage Theatre in Chicago for installation and full-scale testing. Glenda Jensen, then a secretary in Todd's New York office, recalls that Laube was intimately involved in planning the film. He met regularly with Mike junior and scriptwriters William and Audrey Roos in the spring and summer of 1958, crafting a script that would showcase his scent effects. United Artists, which had distributed *80 Days*, agreed to underwrite the film. The widowed Elizabeth Taylor was cast to play the woman at the heart of the mystery in a ten-second-long, smellable cameo.

At the end of the summer, *Film Daily* reported that a public-relations executive named Charles Weiss was planning his own scented feature film. The Weiss Screen-Scent Corp. had lined up Rhodia, a well-known fragrance company, to supply smells to be blown over the audience via the theater's air-conditioning system. The paper reported that production would begin on March 26, 1959, and that release was slated for late 1959 in New York, Los Angeles, Chicago, Philadelphia, and Detroit. Nothing was said about a director, producer, stars, or studio. It would have been hard for Todd and Laube to know how credible this threat was.

Todd began filming in Spain on March 30, 1959, and Bill Doll, the Todds' superstar press agent, set to work building a buzz in the media. A story in *Film Daily* revealed the cast, a new title (*Scent of Mystery*), a new name for the process (Smell-O-Vision), and a release date (an August premiere in Chicago). The story ran with a now-famous photo of Mike junior and Laube on either side of the scent generator's mechanical brain. The *Los Angeles Times* disclosed the movie's ad slogan: *"First (1893) they moved, then (1927) they talked, now (1959) they smell."*

Laube, meanwhile, began installing and testing his system at the Cinestage in Chicago. The odors in his machine were contained in a set of forty 400 cc cylinders or "cells." A syringelike pickup nozzle descended into a cell, extracted 2 cc of fragrance, and injected it into a blower. Scented air was carried into the theater through plastic tubing and released from perforated cylinders (eighteen inches long and three quarters of an inch in diameter) mounted on seatbacks.

Laube shuttled between New York and Chicago every week for months; he hated flying, so he took the train seventeen hours each way. Around June, Laube, joined by his close friend and collaborator Bert Good, began long hours of experimentation in the warehouse space at 1700 Broadway. They were there on a daily basis, fine-tuning the delivery of scent to a mocked-up row of theater seats in their makeshift laboratory. Hal Williamson, then a new employee of Todd Productions, remembers that Mike junior was a frequent visitor to the test site. Finally the system was ready to demonstrate to the United Artists brass, including president Robert Benjamin. Elizabeth Taylor, who now owned Todd's estate and was herself an investor in the project, flew in for the evening demonstration. There was a lot at stake, but the studio execs were impressed with the new technology and agreed to continue their support.

Shooting wrapped on July 4 with the production already badly behind schedule. The planned August premiere was pushed back to year's end; Mike junior told the *New York Times* they needed extra time to finalize the sound and scent tracks. Laube worked furiously. Fortunately his second U.S. patent was issued in September; it got him and Mike junior another mention in the papers.

If Smell-O-Vision caught on, they would need to rush production of enough scent generators to equip moviehouses across the country. A deal was struck with Belock Instrument Company, a Long Island defense contractor that supplied guidance and control components for Atlas and Polaris missiles. Belock was seeking consumer applications for its technology, and they agreed to manufacture the scent machines and to provide state-of-the-art eight-channel stereo sound as well. The company featured a photo of a Smell-O-Vision machine in its October 1959 annual report.

The Todd Organization spent nearly $2 million ($14 million in today's money) producing the film, not a trivial amount in 1959 Hollywood. Shooting on multiple locations in Spain was expensive, as was using 70 mm widescreen cameras and eight-channel sound. Established actors like Peter Lorre (famous for his roles in *Casablanca* and *The Maltese Falcon*) came with a high price tag as well.

The Todd Organization also invested in a host of marketing tie-ins. The Schiaparelli company produced a limited-edition *Scent of Mystery* perfume, the same worn by Elizabeth Taylor's character and smelled by moviegoers in the theater. A thirty-page souvenir program to be sold in theaters included a bound-in soft vinyl record. The movie's title song, sung by crooner Eddie Fisher, was released as a 45 rpm single, along with an LP soundtrack album and sheet music. A novelization of the film by screenwriters William and Audrey Roos, illustrated with stills from the movie, was published as a Dell paperback. Press agent Bill Doll prepared and distributed more than forty individually captioned publicity stills to promote the film, and many of them ran in newspapers and national magazines. This level of expense and effort implies that the Smell-O-Vision team wasn't indulging in a cheap gimmick—they expected a serious return on their substantial investment.

A Challenger Appears

On October 17, 1959, the *New York Times* reported that Walter Reade Jr. was "rushing plans to uncork a smell system of his own before Dec. 22, when Mr. Todd's film opens in Chicago." The forty-two-year-old Reade ran a chain of movie theaters and a movie distribution company (Continental Distributing, Inc.) founded by his father. For

$300,000 he had just bought the rights to a previously released Italian travelogue about Red China, which he reedited and dubbed for scent. At a press conference, Reade revealed that his film, now called *Behind the Great Wall*, would use a new process called AromaRama: *"You must breathe it to believe it!"* Most alarming for Todd and Laube, the Reade picture would premiere in New York on December 2, three weeks before Smell-O-Vision's debut in Chicago. Noting that Reade was "obviously rushing to beat Todd's premiere date," *Newsweek* went for the easy pun and declared that "Todd might be beaten by a nose." Thus was born the epic competition between Smell-O-Vision and AromaRama, a duel that *Variety* dubbed "the battle of the smellies."

According to Reade's press kit, AromaRama dispersed smells through the theater's existing air-conditioning ducts with a boost from Freon gas, while an electronic air purifier prevented odor buildup in the auditorium. A battery of premixed scents would last, it was claimed, for twenty-one performances. Installation costs ran from $3,500 to $7,500 per theater.

Detail for detail, Reade's AromaRama was the system announced thirteen months earlier by Charles Weiss, who was now part of the AromaRama team. This raises a question: Had Reade acquired an independent business from Weiss, or had Weiss been a stalking horse for Reade all along?

BEHIND THE GREAT WALL became the first commercially released smellie when it opened at the DeMille Theater in New York on December 2, 1959. That Reade chose a venue directly across the street from Todd's Warner Cinema was either a coincidence or an in-your-face marketing gesture. The premiere was not a particularly classy event; Joan Didion covered it for William F. Buckley's *National Review*:

> The glory that was AromaRama began even before the theater darkened. Outside, a gentleman in a Tartar falconer's costume

strolled about Seventh Avenue with a stuffed falcon on his arm; the lobby crawled with acned, pigtailed youths in coolie hats and usherettes with Maybelline-slanted eyes and rayon-brocade sheath dresses slit past their knees. Except for the inscrutable fact that everybody on the scene at the DeMille was pure Bronx Caucasian, the ambience seemed roughly that of the old honky-tonk International Settlement in San Francisco. Upstairs, tea was poured for the customers "courtesy of Chin and Lee," who were pushing their canned chow mein in conjunction with this Third Wonder of the Entertainment World.

As for the film itself, the opening sequence featuring a sliced orange was a crowd pleaser. The *New York Times* found the other odors to be "neither so clear nor pleasurable." Luz Gunsberg had the same reaction. Her husband, Sheldon Gunsberg, was Reade's assistant and closely involved with AromaRama. She remembers, "When the film started . . . in the little prolog, he cut an orange and that was incredible. That was fabulous—just wonderful. But after that the smells got all mixed up and they couldn't get them out; so it was a terrible situation." The odors that poured from the overhead ventilation ducts were potent. *Time* magazine reported that they were "strong enough to give a bloodhound a headache," and *The New Yorker* called the experience "quite a massive assault on the olfactory nerves." Says Gunsberg, "my husband would come home and we would have to hang his suits all over the house and open all the windows because we couldn't get the smell out. It really permeated the whole place." Todd employee Hal Williamson bought a ticket to scope out the competition: "Your clothes reeked when you came out of this stuff that had been dumped into the air conditioning system. As I recall there was even a fine mist in the air."

The smells, created by Rhodia perfumer Selma Weidenfeld, were criticized for a lack of subtlety. *Time* thought they "will probably seem phony, even to the average uneducated nose. A beautiful old pine grove in Peking, for instance, smells rather like a subway rest room on disinfectant day." (I sympathize with Weidenfeld; a formula that smells great on a test blotter can fall apart completely when it fills an entire room. Asking her to design at her desk fragrances meant to be smelled throughout an auditorium was like expecting the guy who etches your name on a rice grain to do it in skywriting.) The sheer number and range of the AromaRama smells were overwhelming: jasmine, grassland, incense, spices, soy sauce, a tiger, and a pungent waterfront, among others. Instead of heightening reality, the smells were distracting, according to the mass of critics at the *New York Times, Variety,* and *The New Yorker.*

Then there was the problem of synchronization. Every so often, said *Variety,* "the machine-made olfactory flavors don't correspond with what's on view." *Time* complained that "the smells are not always removed as rapidly as the scene requires: at one point the audience distinctly smells grass in the middle of the Gobi Desert." Paul Baise, who worked for Reade in advertising and public relations, experienced this firsthand. He tells me that AromaRama "worked part-time but not over a period of time, because after a while all the smells melded into one, they overlapped into each other, and they were coming out onto the screen with the wrong image. It was doomed because it got off sync."

More than a whiff of cynicism hovered over Reade's project, beginning with its name: AromaRama made fun of Michael Todd's Cinerama. In the only original footage he added to the movie, Reade took a swipe at Lowell Thomas's introductory appearance in *This Is Cinerama.* In the opening sequence of *Great Wall,* Reade had NBC television news anchor Chet Huntley demonstrate AromaRama by slicing the orange in half. The choice of *Great Wall* as a movie vehicle was another dig at the Todds; travelogues were a Cinerama specialty: *Cinerama Holiday* (1955) and *Cinerama South Seas Adventure* (1958), for example. Reade's tactics got under Mike junior's skin. On his Christmas card for 1959, he printed a verse that began, "Let kind oblivion overtake / all other 'scopes and 'ramas,'" and continued, "Into this world of much dissension / I bring you some fun in a brand new dimension."

Reade's actions were not those of a man expecting great success; according to *Variety,* he made only enough prints to show the film in six theaters simultaneously. He produced no ancillary merchandise, and his openings had none of the celebrity buzz that Todd's did. He didn't bother to incorporate AromaRama Industries, Inc., until one week before the picture opened. Reade promoted his smell system harder than his movie, printing AROMARAMA in gigantic letters atop the ads, with the movie name below it in letters a quarter the size. (The Smell-O-Vision tagline appeared in smaller letters below the title.)

The largely negative reaction to *Great Wall* threatened to spoil the upcoming release of *Scent of Mystery. Variety* noted that Aroma-Rama's New York ticket sales were good but not great, and that Reade's people "apparently aren't expecting any overwhelming jubilation on the part of the trade." *Variety* was prepared to dismiss the idea of "smellies" before Smell-O-Vision had even opened. When I asked him about Reade and Weiss's impact on Smell-O-Vision, Hal Williamson said, "in retrospect they probably did more to harm our cause than the occasional failure of [our] scents to work exactly as they were supposed to. It left a very bad taste with the press after the Reade opening in New York." Even Reade's people admit to the problems. Paul Baise says it "was doomed before it even got off the ground, but we went ahead with it anyway and presented it as a piece of new innovation." AromaRama, he says, "belonged in the laboratories, and not presented to a paying public."

Todd Junior Fights Back

Scent of Mystery premiered in Chicago on January 12, 1960, with all the hype the formidable Todd PR machine could provide. A chartered plane flew Elizabeth Taylor in from New York, accompanied by members of the press. The producers threw a preshow cocktail party at Fritzl's, a showbiz watering hole. The film was preceded by *The Tale of Old Whiff*, a cartoon with fifteen Smell-O-Vision scents and Bert Lahr (of Cowardly Lion fame) as a character voice. At a late dinner following the movie attended by nearly 250 people, the entertainment included Milton Berle, Henny Youngman, and Mort Sahl. Cohosting the event with Todd junior was Elizabeth Taylor, recently married to Todd senior's showbiz buddy Eddie Fisher. At the New York opening on February 18, Taylor's presence drew a huge crowd of fans and reporters.

The film itself was received warmly, if not enthusiastically. Most critics liked the exotic scenery and action sequences. *Variety*'s take was typical: "Diverting tale told with nostril-appeal." The *New York Times*'s Bosley Crowther was the rare critic who disliked the film itself, from the "whole silly plot" to the acting ("downright atrocious" and "virtually amateur"). As for the smells, Crowther seemed to have trouble getting them; he said they were "the least impressive or even detectable features of the show"; every so often, he detected something "faint and fleeting."

The Smell-O-Vision scents played off the screen action in clever ways. When Peter Lorre's character drank coffee, the audience smelled the brandy in it. When Denholm Elliott slipped and almost fell in an outdoor market, the audience smelled (but didn't see) a banana—an aromatic twist on a very old sight gag. Topping it all off, the smoke from Peter Lorre's pipe holds the key to the plot's mystery.

Who Won?

Comparing Smell-O-Vision to AromaRama, Hollis Alpert, writing in *Saturday Review*, was even-handed but unsympathetic, saying that "neither is particularly successful or desirable. Differ though they may in technology, the smells are equally synthetic, and equally erratic." Most other reviewers gave Smell-O-Vision the edge in aesthetics. *Time* said its odors were "on the whole no more accurate or credible than those employed by AromaRama, but at least they don't stink so loud." According to *Variety*, "The Smell-O-Vision odors seemed more distinct and recognizable and did not appear to linger as long as those in AromaRama." *The New Yorker*'s John McCarten said, "After a lot of thoughtful recollective sniffing, I should say that Glorious Smell-O-Vision is subtler than AromaRama. Professor Laube seems to have mastered the quick change; in any case, he is able to get the smell of coffee out of the place before the loaf of fresh bread appears on the screen."

But it wasn't just Laube's efforts that gave Smell-O-Vision its edge. Many years later, Mike Todd Jr. credited his press agent Bill Doll with the idea of reversing the odor pump after each delivery to reduce lingering of previous smell. "Bill got this idea after the third opening. It was used, and it worked perfectly, but by that time the ship had sailed."

Back in 1939, when he was promoting Odorated Talking Pictures, Hans Laube had said ten smells would suffice for a feature-length film, because more would be "too much for the public's nose." In his 1956 patent application, Laube increased the optimal number to between twelve and twenty. *Scent of Mystery* was released with thirty. In the competition to show off their new systems, both Todd and Reade had oversaturated their audience.

A QUESTION OF personality lingers over the battle of the smellies. Mike Todd Jr. had little of his father's fire. He was polite and tentative. Anticipating ridicule, he adopted a tongue-in-cheek attitude toward Smell-O-Vision that signaled a lack of seriousness to critics and distributors. The movie critic Hollis Alpert Jr. found him "a somewhat timid revolutionist."

The elder Todd took great pleasure in gambling on his own talents. According to his son, "He was at his best when the odds were against him and a show was in trouble and he needed to utilize all of his energy and ingenuity." Todd senior was strongest late in the game. His contribution to a show began after rehearsals; he switched into top gear only during out-of-town preview performances. "He thought best on his feet, under pressure," said his son. He was legendary for last-minute adjustments to shows and promotions that made winners out of questionable properties. And not least, he was a great motivator of other people: he knew how to drive technical wizards to produce workable, show-worthy effects.

One is tempted to ask: Would Smell-O-Vision have taken off if Michael Todd Sr. had lived? It is easy to imagine him pushing perfumers to the limit, stalking about the floor of the Cinestage before opening night to tweak the scent delivery. Todd senior's showbiz sense would have kicked in; the film would have been snappier and the scent effects more polished. His genius for promotion would have taken flight—imagine him pushing *Scent of Mystery* perfume with the help of his glamorous movie star wife. He would have schmoozed the stuffy-nosed Bosley Crowther and his colleagues in the press. Above all, he would have reacted quickly to Reade's tactics, and maybe played them to his advantage.

Hal Williamson says, "if we could have survived another couple of months probably, the fine-tuning could have been done. But at that point the critical and public reactions were such that Michael and Elizabeth decided not to keep going with it."

Smell-O-Vision—its technology, its film, and its promoters— was a serious entertainment gamble, even if it was a long shot. AromaRama, in contrast, never had any legs at all. Technologically, business-wise, and aesthetically, it was a cynical rabbit punch of counterpromotion. Smell-O-Vision was more than a gimmick, but

AromaRama was something less, a mean-spirited exploitation. Walter Reade ambushed Mike Todd Jr., then dogged his every turn. Temperamentally unsuited for the rough-and-tumble of showbiz, Todd gave his more aggressive rival too much room to maneuver. Although critical opinion tilted toward Smell-O-Vision, Reade had effectively killed any prospects for its commercial success.

WERE SCENTED MOVIES simply gimmicks? John Waters thinks so. He tells me that his inspiration for Odorama was William Castle, whose promotions in the 1950s were the very definition of Hollywood gimmicks. Castle, for example, hid vibrating electric motors under random seats and set them off during the Vincent Price horror film *The Tingler.* Castle's stunts were cheap and easy—no inventors spent long years in the lab perfecting them, and no lawyers were paid to file patents, incorporate companies, and draw up licensing agreements.

I ask Waters if movie smells can be anything other than a gimmick. "You mean for real in a drama? No. I think it will always be a gimmick, because it takes you out of the movie.

"To me, what made *Polyester* work were bad smells. All the movies had good smells. We started with a good smell, and ended with a good smell, but we had bad smells all through it and that's what made it successful. Never is it going to be successful if it's good smells. It's boring. You have bad ones, it's funny. If it's ever used again, it will always be for comedy."

But despite his protestations that it's all in good fun, when the *Rugrats Go Wild* feature-length cartoon came out in 2003 with scratch-and-sniff "Odorama" cards, John Waters hit the roof. Attorneys for his studio, New Line Cinema, went to work, and in short order the Rugrats and their corporate owners at Nickelodeon and Viacom dropped the use of the name Odorama.

At the heart of every gimmick is an idea worth defending. The notion of scented entertainment—whether in the movies, a dance club, an opera, or a concert hall—remains attractive and widely popular. As an added dimension, it offers all the possibilities of sight and sound: compelling realism, surprise, and emotional transport, as well as sly commentary, comedy, and ironic distance. I have no doubt that a director with sufficient olfactory genius could create a superbly entertaining smellie. It's unfair to ask such a person to develop the necessary technology as well. Somewhere in our wireless and digital world there is an elegant way to deliver scent to an audience. When it becomes a reality and falls into the right creative hands, we may see a new dawn of Smell-O-Vision.

Aftermath

The golden age of scented movies was brief but spectacular. It began in the spring of 1958 and was over by the summer of 1960. Neither Smell-O-Vision nor AromaRama would ever be used again.

The equipment Reade used for AromaRama—whatever it may have looked like—has vanished. When Mike Todd's Cinestage Theatre in Chicago was about to be gutted in 1994, cinema buff Marc Gulbrandsen sneaked in to take a last look around. He spotted the old Smell-O-Vision equipment in the basement, but it was never recovered.

Carmen Laube, the daughter of the man who invented Smell-O-Vision, has an apartment on the Upper West Side of Manhattan. Her father was fifty-six years old when she was born, so she is too young to remember his excitement about *Scent of Mystery*. She does remember his passion for scent, and the disappointment of his old age when his entrepreneurial spirit waned at last. She showed me photographs of her father. He is dapperly dressed and always wears his signature dark-framed eyeglasses. The snapshots are from the deep past: Laube behind the wheel of a racing car in Switzerland in the 1930s, in a dinner jacket on board the luxury liner *Andrea Doria,* and finally at the 1939–40 New York World's Fair, standing next to the packing crates that carried the Odorated Talking Pictures equipment from Zürich to Flushing Meadow for the screening of *My Dream.*

Carmen opens a box of memorabilia and hands me tickets and an invitation to the Chicago premiere of *Scent of Mystery:* "Mrs. Eddie Fisher and Mr. Michael Todd, Jr. take pleasure in inviting you . . ." There is the printed menu from the post-film supper party—a glamorous midnight affair at the Ambassador West Hotel, with two bands and "impromptu entertainment by our friends from the world of show business." There is the neatly folded stock certificate embossed with a corporate seal: 200 shares of Scentovision, Inc., to Hans Laube.

I speak on the phone to Hans Laube's widow, Novia, who now lives in Florida. Through her heavy Estonian accent I hear fierce determination and loyalty. She tells me how she met and married this tall, handsome, intellectual European; how particular he was about his clothes—the fine suits and custom-made shirts. How hard he worked, often late at night, and about the seven months he spent commuting to Chicago to prepare Smell-O-Vision for its debut. For the Laubes, a lot was riding on Smell-O-Vision. She tells me, "Michael Todd and everybody said the name Laube would be known all over the world. Because we anticipated that this would be a great success."

When I ask about the competition with Walter Reade and Aroma-Rama, her tone sharpens. "He came out just a few weeks before us, or just a month before us. He spoiled the entire idea because when people went to see his movie the smell clung to their clothes and they said, 'Oh no, no, we don't want that.' . . . [Reade] wanted to make money, he wanted to come out before us, and he stole my husband's idea." The failure of Smell-O-Vision was a financial blow to Laube. Novia says Michael Todd promised her husband a nickel for every ticket sold. The film ran for months, but "they did not give Hans one single penny. So that was a terrible disappointment too. They did not keep their promise." It took a psychological toll as well. "It killed my husband mentally," she says.

After the movie closed, Laube rented laboratory space on East Eighty-fourth Street, where he developed an electronic home fragrancer called the Bestair, but the device was ahead of its time and never made it to market. The organizers of the U.S. exhibit at the 1964 World's Fair approached him about a scented movie project, but dropped it at the last minute. With that final, crushing disappointment, Laube threw in the towel. "I had to take care of my husband for twelve long years . . . to support him after that because he ended up penniless, totally penniless." After years of declining health, Hans Laube died in 1976, at the age of seventy-six.

IN THE CORNER of Carmen Laube's living room, topped by a collection of ornate table lamps, sits a shiny stainless-steel cabinet. Behind its clear Plexiglas face I see motors, pumps, gauges, and dials, and above them a turntable ringed with glass bottles. I'm looking at the ultimate Smell-O-Vision artifact, the working prototype her father used to fine-tune scents for Mike Todd's movie forty-seven years ago. A lever arm above one flask is frozen in place like the Tin Woodman's arm, forever poised to descend and extract the next scent. The smell has long since evaporated.

CHAPTER 9

Zombies at the Mall

All around the world people and companies are becoming
aware of the power of scent.

—MARTIN LINDSTROM, *Brand Sense*

NASAL PERSUASION IS HAPPENING EVERYWHERE. A
scent generator hidden in a ventilation duct, or parked dis-
creetly in the corner, can amplify the natural scent of a
store's merchandise: high-end shirtmaker Thomas Pink plays
freshly laundered linen, while the Hershey's outlet in Times Square
vents extra chocolate into the air. Some merchants get creative, like
the furniture store in Massachusetts that filled its children's section
with a bubble-gum scent. Even brands with no inherent scent get in
on the act: consumer electronics giant Samsung wafts a corporate
logoscent into its flagship store on Columbus Circle, and Westin
Hotels uses a signature "White Tea" composition in the lobby. In
each case, by providing a more engaging retail experience, the company hopes to benefit in terms of sales, consumer satisfaction, and
brand image.

Are we on the brink of a new era in advertising? Marketing
wunderkind Martin Lindstrom believes so. In his recent book
Brand Sense, extolling the future of multisensory branding, Lindstrom is extra-super-excited about scent—he sees it as the next huge

trend in marketing. Whether or not scent becomes an integral part of branding, Lindstrom's enthusiastic prediction is the latest in a long history of marketing to the nose.

In 1925, for example, a headline in New York's *Daily News Record* read "Sense of Smell—An Important Factor in All Modern Merchandising." In 1934 *Forbes* told its readers, " 'Sell by Smell' may be the next big slogan in marketing." In 1939 *The Management Review* said, "The odor engineer is joining the color engineer as a consultant to the sales manager." In 1947 *The Saturday Evening Post* warned that "Shrewd merchandisers have charted a new route to your pocketbook. Now, shoe polish smells like roses, ink is perfumed, imitation leather has the scent of pigskin."

Today's merchandisers continue to experiment, with such offerings as lavender-scented automobile tires (aimed at women) and high-end bowling balls redolent of orange-ginger. The real action, however, lies in projecting olfactory character into indoor commercial spaces. This application has been fully embraced in one large business sector: the gaming industry. Las Vegas is the trend's epicenter; half the major properties on the Strip have scent systems. The MGM Grand has deployed as many as nine scents simultaneously around its property, and the Venetian features a corporate logoscent called "Seduction." In their quest to fine-tune consumer experience, casinos have made sensory engineering a priority. Guest rooms are kept chilly to discourage visitors from spending too much time in them. Complex floor plans channel patrons farther into the gaming areas, where clocks are banished, along with views of the outside world. In seeking new ways to keep people playing longer, casinos have taken the lead in manipulating the commercial smellscape.

A negative example—the removal of a brand's characteristic smell—reveals the importance of the olfactory dimension. As the Starbucks Coffee chain expanded, it decided to switch from open containers and store-ground coffee to flavor-locked packaging. Its goal was to ensure the freshness of its roasted beans and to make life easier for the java-jockeys. But the vacuum-sealed packaging came with an unanticipated cost: it made the shops aromatically sterile. Without a coffee-heavy atmosphere to entice them, customers were being poached by the competition. Starbucks lost what company founder Howard Schultz calls "perhaps the most powerful non-verbal signal we had." To get it back, the chain is considering a return to scooping and grinding actual beans.

Businesses can be confronted with olfactory issues by a sudden change in public policy. When smoking in pubs and clubs was recently banned in Scotland and Wales, owners were shocked to discover how bad their establishments smelled. Once the smoke cleared, Luminar, a company that owns a chain of British night-clubs, found that "the stench of beer and sweat was no longer masked by smoke." The company began a frantic search for ways to mask the unpleasant new reality. The proposed remedy—blowing rose scent over a mass of sweaty, burping bodies—doesn't sound promising, but here's hoping they find an effective solution. Fraternity houses across America will be paying attention.

A DECADE AGO, the social psychologist Robert Baron cased a shopping mall near Albany, New York, mapping out odorless areas as well as spots that had a naturally pleasant scent—the latter turned out to be near Mrs. Field's Cookies, the Cinnabon store, and The Coffee Beanery. Next, Baron sent in accomplices who approached shoppers and "accidentally" dropped a pen or asked for change for a dollar. Baron recorded one simple response: Did the shopper help the stranger or not? Helping behavior—picking up the pen or making change—was significantly higher in pleasantly scented areas than in unscented ones. Baron's experiment was the first to examine the effects of odor outside the lab and in a natural consumer ecosystem—the mall. Its result was clear: shoppers respond to ambient scent in measurable and meaningful ways. The familiar scents of daily life may not call attention to themselves, yet they exert subtle behavioral effects on those who inhale. Did Cinnabon set out to make mall patrons more helpful? Unlikely, but it turns out helpfulness is just a side-benefit of public bun baking.

The Albany mall study whetted the appetite of psychologically inclined marketers. They wanted to know if scent could have more useful, or more profitable, effects on consumers. They wanted scientific evidence that scent could sell, and most of all, they wanted to know how it worked. With few exceptions, like Baron's study, the scientific exploration of those questions takes place in psychology labs, with college sophomores as stand-ins for regular consumers. In a typical arrangement, students are brought into a room and asked to rate images of products on a computer screen, or to evaluate merchandise in a mocked-up store display. Sometimes the room is scented, sometimes not. Generally, researchers find that scent can change attitudes toward merchandise, but it's risky to extrapolate from such highly contrived experiments to real-world uses. Research continues, however, and marketers forge ahead, even without the imprimatur of science.

So how does a scent in the air change behavior? From the literature of social psychology, Professor Baron knew that positive events gave rise to small and brief improvements in people's moods. Something as trivial as finding a coin in a pay phone will do the trick. (The coin finder, for example, is more likely to agree to take part in a boring task a few minutes later.) Baron reasoned that the aroma of coffee and baked goods made people more helpful by lifting their mood. Sure enough, follow-up interviews revealed that shoppers in the scented areas were measurably happier than those in unscented areas.

Baron's mood hypothesis was easy for marketers to accept because it closely resembled the conventional wisdom that smell was a purely emotional sense. This means that scent marketing is mood marketing; and creating mood is something marketers feel they understand. The equation is simple: nice scent equals good mood equals increased sales. Baron's explanation also appealed to professional vanity: it cast the spritzer-wielding marketer as a voodoo priest, able to pull the scent-addled public through a store like the iron filings on a Wooly Willy. Mood theory became the rally cry of scent marketers everywhere. The senior PR director for Westin Hotels & Resorts, and the woman behind their White Tea logoscent, subscribes to it. "We wanted to make an emotional connection," she says.

THE NOTION THAT smell is purely an emotional sense is an old one. In 1924 the chemist and physicist E. E. Free, a former editor of *Scientific American,* said, "Practically all the reactions to smells are emotional effects on the part of our mind that is called 'unconscious.' They are not reasonable, intellectual reactions at all." Free backed up his claim with a bizarre anecdote about a man who became unaccountably angry whenever he smelled horseradish. Today scientists continue to offer sound bites about the emotional force of smells. The social anthropologist Kate Fox tells the BBC, "Our sense of smell is directly connected to our emotions," and "Smells trigger very powerful and deep-seated emotional responses." The German psychologist Bettina Pause says, "Odors seem to be powerful emotional stimuli." The English psychologist Steve van Toller tells *The Independent,* "Smells plug straight into our emotional centres in the middle part of the brain—the nonverbal part—and can have a powerful effect on our feelings." The American psychologist Rachel Herz explains to *The Lancet* that the nose "has direct access to the amygdala," the portion of the limbic brain that controls emotional response. Quotes like these set a marketing manager's hair on fire. Who wouldn't want to plug their brand straight into the emotional center of the brain?

Alas, things are not that simple. A big challenge to the mood theory of scent marketing is the "congruency" problem. Studies repeatedly find that for a scent to be effective, it must match its commercial context. A mismatch produces no benefit, and may even leave consumers with an unfavorable impression of a store or brand. For example, one experiment used two equally pleasant fragrances: *Lily of the Valley* and *Sea Mist*. One or the other was in the air as female college students were shown a display of satin sleepwear for women. The students said they were more likely to purchase the clothes, and were willing to pay more for them, when *Lily of the Valley* was in the air. In separate testing, *Lily of the Valley* was rated as a better match to the clothes. While *Sea Mist* was equally pleasant, it lacked the feminine associations and bedroom ambience of *Lily of the Valley*. So much for nice scent equals good mood equals increased sales—people pay attention to the meaning of smells.

The congruency problem popped up again when researchers examined the combined effects of ambient music and scent in an actual gift store. They played tunes that were either relaxing or energizing, and used scents with either high or low arousal value. When low-arousal lavender was paired with relaxing tunes, the result was a significant increase in consumer satisfaction and impulse purchasing, and a higher interest in exploring the store and making a return visit. The same happened when high-arousal grapefruit was paired with energizing tunes. Yet the same tunes and smells, when mismatched for energy level, had no effect on consumer behavior. In another study, photos of a store decorated for a holiday sale got favorable ratings when shown with a Christmas-themed fragrance and Christmas-themed tunes. The photos got lower ratings when the Christmas scent was paired with nonholiday music. The overall lesson is clear: for smell to be effective in marketing, context matters, because people try to intellectually reconcile what they see with what they smell.

The market is already addressing the need for multisensory coordination. Retailers who don't have the time or skill to invent their own blends of scent and sound can select from prepackaged combinations. Muzak LLC, the company that supplies background music for stores and offices, has teamed up with Scent-Air Technologies, Inc., an outfit that installs aroma equipment in retail stores. Together they offer custom-designed scent-and-sound combinations that "enhance the retail experience." Scent-Air's CEO told newspapers, "We're Muzak for your nose."

Recently, the University of Washington business professor Eric Spangenberg and his colleagues gave marketers just the kind of study they'd been yearning for: one that measured the effect of scent in dollars and cents. Spangenberg's team used an actual off-campus clothing store, where half the floor space was devoted to men's clothing and the other half to women's. Over the course of two weeks the store was alternately scented with two fragrances of similar strength and pleasantness: a feminine vanilla and a masculine rose maroc (a spicy, honeylike note). When vanilla was in the air, women's-wear sales increased and menswear sales declined. When rose maroc was used, the sales changes were reversed. In other words, men bought more when the scent was male-appropriate, and less when it was feminine; the reverse was true for women. The effect was substantial. People shopping under gender-appropriate scent bought an average of 1.7 items and spent $55.14; people shopping with the gender-inappropriate scent bought only 0.9 items and spent $23.01.

Call it congruence or call it context—the important point is that the judgments affected by scent involve comparison and evaluation, not just an emotional gut-check on the part of the consumer. The shopper who perceives a mismatch between a store's scent and its goods or music is using reasoning, not feelings. The narrow focus on emotion is beginning to give way as more researchers find that consumers process smell information cognitively. Marketing experts are beginning to give people credit for thinking. The Canadian researchers Jean-Charles Chebet and Richard Michon, for example, believe that emotion has been overemphasized as an explanation. They manipulated the scent of a mall near Montreal and found that mood had relatively little impact on how much shoppers bought. Chebet and Michon contend that scent instead changes how shoppers think about the appearance of the mall and the quality of its merchandise. In other words, what counts is meaning more than mood.

Once outside the psych lab, the concept of congruency doesn't offer marketers much traction. Academics know congruency when they see it, but they have a hard time explaining in practical terms how a fragrance matches its marketing theme. Out in the real world, fitting a scent to a commercial context has always been a matter of style, taste, and culture. It's what perfumers and fragrance evaluators do for a living, and marketers are well advised to join forces with these experts. What marketers need to do is develop clear standards for success. For example, is the point of a scent campaign to encourage people to stay in a store longer, perceive the goods as trendier, or try a new product? Once a program is under way, it would be useful to have a way of measuring its effectiveness: one can imagine standardized measures of scent delivery (the number of noses stimulated) and effectiveness (e.g., increase in brand awareness). In short, marketers need a Nielsen rating for the nostrils.

DEEP IN THE hair-care aisle of a supermarket, a shopper pops the top on a shampoo bottle and takes a sniff. What happens next is a cascade of decision-making: Does it smell too feminine? Is it refreshing, as the packaging claims? Does it smell like an effective antidandruff product? Will my spouse like it? Does it smell classy enough to justify the higher price? All these questions are asked and answered in two sniffs. To the casual observer, the shampoo sniffer is making a snap judgment—nothing more than an emotional reflex of "do I like it?" Yet in that brief moment, fragrance speaks to status (elegant, cheap, old-fashioned), functionality (cleansing, conditioning, therapeutic), and self-identity (feminine, edgy, safe). The scent is full of information, and the consumer is analyzing it. Fragrance speaks to the emotions, but it is more than mood music. It can carry a message to the mind. Once marketers master this sophisticated language, the sense of smell will become a full-fledged advertising medium.

Subliminal Scents

Any marketer who thinks of using smell wants to know how it works, so that he can build a strategy to take advantage of it. Conventional wisdom, slow to acknowledge new research results, still emphasizes emotion as the main psychological mechanism, and thus marketers continue to select scents based on their emotion-inducing qualities. But deciding how strong or weak to set the aroma level is a different issue, one that inevitably leads to questions about the nature of conscious awareness.

No topic in psychology fires the popular imagination as surely as subliminal perception. The mere phrase evokes (subliminally!) technicians in lab coats twiddling dials on a control panel as consumers sleepwalk to the checkout line with armloads of unwanted merchandise. Can a secret scent really turn us into zombie shoppers? Can we be made slaves to smell?

To a psychologist, *subliminal* has a fairly dry technical definition; it means "below the threshold of conscious awareness." A subliminal stimulus is too weak to be perceived with certainty, yet strong enough to leave a brief, featherlight impression on the senses. These faint and fleeting perceptions, which elude the direct gaze of our attention, cannot be measured by the traditional methods of rating scales and adjective checklists. Instead, they must be measured by their indirect effects on other mental processes. For example, one can flash "DOG" onto a screen so quickly that a viewer has no time to read it, and can't even be sure he saw anything. It is pointless to ask him to identify the word. Yet the flashed word causes a flicker of measurable brain-wave activity, and its lingering trace will be evident in subsequent word-association tests.

It is a deeply held belief of marketers that scented advertising works subliminally. For example, according to Sue Brush, senior vice president of Westin Hotels & Resorts, the chain's *White Tea* fra-

grance is "one of those subliminal things you don't necessarily adver-
tise, but we hope it can help guests decompress after the rigors of the
road." Enthusiasts and detractors both believe that scent marketing
is a form of mind control that operates in the murky zone of the subliminal, where a well-placed whisper is all that's required to set off
psychological chain reaction resulting, inevitably, in an opening of
the consumer's wallet.

ACCORDING TO THE PSYCHOLOGIST Anthony Pratkanis, popular
enthusiasm for the subliminal has come in waves. The first arrived in
1957, when James Vicary claimed to have shown subliminal ads in a
movie theater. Vicary said his messages—"eat popcorn" and "drink
Coca-Cola"—boosted Coke sales in the lobby by 18.1 percent and
popcorn sales by 57.7 percent. In the Cold War era preoccupied with
the brainwashing of soldiers and secret agents, Vicary's claim generated enormous media coverage. Yet Vicary couldn't or wouldn't produce his data. Nor would he show anyone the tachistoscope he
allegedly used to flash the ads onto the movie screen. He eventually
admitted to *Advertising Age* that he fabricated the study to draw
attention to his consulting business.

A second subliminal wave began in 1973, when Wilson B. Key
published *Subliminal Seduction,* in which he claimed that sexually
arousing images were hidden in printed advertisements. (This led to
a brief fad at parties in the mid-1970s, where people squinted at
whiskey ads in *Esquire,* looking for a sex orgy in the ice cube.)
The original studies cited by Key were flimsy and lacked critical
control groups. Though his theories were roundly dismissed by
psychologists, Key—now an elderly man—continues to see penises
embedded in advertising images wherever he looks.

The third and most recent wave of the subliminal fad came in the
late 1980s and early 1990s with self-help audio tapes that promised
everything from weight reduction to increased self-esteem. Driven

partly by late-night infomercials, subliminal tapes became a $50-million industry, even though little or no scientific evidence existed that they worked as claimed.

It's clear that we can absorb visual and auditory information without being consciously aware of it. Whether these fleeting perceptions affect our behavior as directly and purposefully as subliminal-advertising proponents claim is another story. Anthony Pratkanis finds no evidence that they do. I believe the same holds true for smell. There is, for example, solid evidence for subliminal odor perception. The German researcher Thomas Hummel snaked a millimeter-wide tube about three inches up the noses of volunteers. (Actually, he let them do it themselves—it's less stressful.) The tube delivered a constant stream of warmed and humidified air, along with occasional pulses of odor, directly to the sensory surface of the nose. A wire inside the tube monitored electrical activity from the same surface. Scents too weak to be consciously detected nevertheless provoked a response in the sensory cells of the nose. Using different techniques, other researchers have observed the brain responding to scent at levels too low for the test subject to reliably detect. There is little question that odors can be registered subconsciously in the nose and the brain.

Psychologists in the Netherlands took techniques used to measure the indirect effects of subliminal sights and sounds and applied them to olfaction. They gave people an incidental exposure to the citrus scent of a familiar all-purpose cleanser. Most participants were unaware of the smell and of the purpose of the experiment. Yet those who inhaled the scent were faster at picking out cleaning-related words from a list, and were more likely to mention cleaning-related behaviors when asked to describe their routine daily activities. Given a crumbly cracker to eat, people who'd been exposed to the cleanser scent engaged in more crumb-sweeping and other tidying behavior than people who hadn't been exposed. The subliminal scent activated a mental network of cleaning-related associations, later expressed through word and deed, but not in a readily exploitable way. People didn't spontaneously mention brand names or rush out to buy a bottle of cleanser. Enhanced crumb-brushing is hardly the stuff of mind control.

That the nose and brain respond to subliminal smells under ultraprecise laboratory conditions is not surprising, but are the effects robust enough to make a difference in the real world? The classic demonstration of covert selling power dates back to 1932. Donald Laird had male students at Colgate University pose as market researchers and go door-to-door in Utica, New York. The young men presented housewives with four samples of identical silk stockings and asked them point to their favorite pair. The stockings varied only in smell: the unadorned product had a slightly rancid character; the others were lightly scented with either narcissus, a fruity note, or a sachet fragrance. Laird's team completed 250 interviews before one suspicious lady called the cops, and when the police report made the local newspapers, the study's cover was blown. Of the 250 women, only six were aware that the stockings were scented. Despite this, there was a clear influence of scent on stocking preference: 50 percent of the women chose the narcissus-scented pair, 24 percent chose the fruity pair, 18 percent chose the sachet scent, and the natural hose were selected by only 8 percent.

Smell alters our behavior in daily life, in the trivial sense that a whiff near lunchtime may steer us toward a burrito instead of a pizza. The subliminality of the message—whether I smell a pizza before I have a conscious desire to buy one—is of no more consequence than whether I heard a pizza ad on my commute that morning. In either case, the compulsion—or lack thereof—is about the same.

Still, subliminal advertising continues to frighten people who should know better. The European Chemoreception Research Organization, a society of smell and taste researchers, recently editorialized about a study done by some of its members, in which smells were presented along with odor-evocative words. The results: people found a cheesy aroma less unpleasant when it was paired with the phrase "cheddar cheese" than when it was paired with "body odor." The power of suggestion was so strong that people reported that even clean air smelled bad when labeled "body odor." This entirely predictable outcome was enough for ECRO to raise an alarm: "Unfortunately this fact offers powerful tools for manipulating the information and directing the choice of consumers towards particular foods, perfumes, [and] detergents," a possibility that, "disturbingly," could lead to "misleading messages." Shocker! Ads seek to manipulate consumer choice. EU bureaucrats will have a field day drafting regulations banning smell fraud in advertising.

Contrary to popular belief, the Federal Communications Commission has no formal rules about subliminal advertising, smelly or otherwise. In fact, the FCC has investigated only one complaint about subliminal messages. In 1987 it found that Dallas radio station KMEZ-FM broadcast a program containing them. Which dastardly corporation was responsible for this outrage? Well . . . none, actually. The subliminals were hidden in an antismoking program aired on behalf of the American Cancer Society.

The idea of subliminal advertising continues to haunt the field. Merchants who use ambient scent are reluctant to talk about it because they don't want the public to view them as zombie masters. They could defuse the issue by debunking the power of subliminals, but they don't—perhaps because they too believe in it, if only a little bit. Subliminal perception is now something experts debate as they recommend fragrance levels for their retail clients. Michelle Harper, director of fragrance development at Ayrlessence, says, "You want it to be subliminal, especially in an environmental space." On the other hand, Joe Faranda, chief marketing officer for International Flavors & Fragrances, says, "The scent no longer has to be working subliminally to be effective." Who to believe? In my experience, when a scent calls attention to itself, people feel obliged to decide whether or not they like it. At that point they're focused on the scent and not the store. Samsung's corporate logoscent—suggestive of green melon—works because it is barely detectable; any stronger and customers would start looking for the fruit salad bar. There's a difference between subtle and subliminal.

Rage Against the Machine

When the English perfumer Eugene Rimmel created the first mass-marketed perfumes in the mid-1800s, he also invented various ways of promoting them through scented print advertising. He gave away scented almanacs and scented fans. He placed scented ads in London theatre programs. These efforts were not met with universal applause. His sophisticated contemporaries turned up their noses at the theatre programs; these aromatic momentos of "rank commercialism" were seen as intrusive, crass, and annoying. The equivalent in our time are scented perfume ads in magazines. Calvin Trillin has inveighed against the ones he found in *Vanity Fair*; they "revived old thoughts about whether the Drafters could have envisioned the possibility that the freedom of expression guaranteed in the First Amendment would someday extend to smelling up the place."

The scented ads that offend Mr. Trillin are the legacy of Fred and Gale Hayman, the California entrepreneurs who started the Giorgio of Beverly Hills boutique on Rodeo Drive. In 1982 they launched a marketing campaign for a perfume named after their store. They began by mailing perfume-soaked blotters to their local clients, but to get samples under noses on a national scale they needed a cheaper method. Their ad for *Giorgio,* in the May 1983 *Vogue,* was the first ever to use the ScentStrip Sampler, a new product from Arcade Marketing. This was the now-familiar printed page with a glued-down flap; as the flap is pulled open, microdroplets of fragrance oil in the glue are ruptured and scent is released. Readers complained that the magazine reeked of *Giorgio,* but sales boomed and the magazine industry never looked back. (Determined to reach even more nostrils, the Haymans unleashed the Spritzer Ladies from Hell, teams of white-and-yellow-jacketed reps who aggressively misted millions of people in department stores.) The *Giorgio* perfume, formulated with an extraordinarily high ratio of fragrance oil to alcohol, was brassy, penetrating, and easily recognized. Fancy restaurants banned it, and wearers caused near-riots in elevators across the country. *Giorgio*-bashing became a snob sport. Outside of Le Cirque and the refined precincts of Manhattan's Upper East Side, however, the perfume was a blockbuster.

Scented print ads are enjoying a new renaissance at the moment. Fox-Walden Films recently paid $110,000 to run a scented full-page movie ad in the *Los Angeles Times*. The *Wall Street Journal* and *USA Today* are said to be considering rub-and-smell ads. Each year the annual report of spice maker McCormick & Company features a different aroma; in 2006 a disappointingly thin nutmeg rendition struggled to be noticed above the stink of the ink. A cover of the German scientific journal *Angewandte Chemie* smelled like lily of the valley, in order to draw attention to an article on odor receptors. The core market for scented ads has always been women's fashion magazines; the publisher of *Allure* claims that 85 percent of her readers immediately try the scent strips in her book.

Among some social critics, scented ads inspire violent imagery; words like "assault" and "bombardment" get thrown around. To the journalist Emma Cook, consumers are helpless victims: "Whereas you can exercise the choice to stop listening or watching, physically you can't help smelling things." Artificial scents put A. S. Byatt, the English novelist, into a foul mood: "I think we are bringing up a generation . . . desensitised by constant loud and garish smells." If man-made scents were sounds, "they would be a cacophony." Byatt is a formidable intellectual who has deconstructed the writings of Wordsworth and Coleridge and lectured on American literature at University College London. How does she account for the inexplicable desire of the masses for scented products? She blames advertising.

"The television screen shows branches and violets. It shows pine forests and sheets of falling white water ending in curls of clean, shining spray. It shows meadows full of buttercups and pine forests full of mystery and crisp needles. It is telling you—enticing you—to recreate these atmospheres in your own home with air fresheners, with aerosol sprays of scented furniture polish, with . . ." You get the drift.

Byatt objects on ethical grounds: "The smells that have invaded our modern lives are neither the good smells nor the bad smells, but the guilty, masking smells. Smells that we use to cover human smells." Apparently perfumes are deceitful because they hide our true primate stinkiness.

Unsurprisingly, Byatt's fiction is riddled with morbid smells. Here's a typical example: "It was not a clean train—the upholstery of their carriage had the dank smell of unwashed trousers." Elsewhere she describes a husband's "evil-smelling breath full of brandy and stale smoke." Occasionally she outdoes herself: "It was a liquid smell of putrefaction, the smell of maggoty things at the bottom of untended dustbins, the smell of blocked drains, and unwashed trousers, mixed with the smell of bad eggs, and of rotten carpets and ancient polluted bedding." Her preoccupation with unwashed trousers gives the impression of a nose tuned to the Dark Side. She recoils from perfume like the Wicked Witch from the fire bucket. Hide the *Giorgio* or she'll send the flying monkeys after you.

Perhaps an elderly British novelist is entitled to get cranky about perfume, but why should a thirtysomething Internet columnist lose it over an air freshener? That's what happened when Mark Morford, of the *San Francisco Chronicle's* SFGate.com website, teed off on Procter & Gamble's ScentStories aroma player:

What vile marketing decision was made, and by whom, that said we must now progress from static mute little tabletop chemical-bomb air fresheners to more sinister, electronically activated Glade plug-in thingies with silly little built-in fans to full-fledged toaster-size appliances that require huge amounts of plastic and massive marketing campaigns and full AC power and interchangeable chemical-soaked disks?

It's not just the ever-grander technology that makes Morford hot under the collar—it's the implied message contained in the aroma:

This is the marketing strategy: each disc is apparently designed to somehow lift you out of your sanitized tract-home suburban kids-'n'-dogs-'n'-minivans dystopia and transport you straight to the Misty Mountains or the sultry Bahamas or the Brazilian rain forest or whatever.

What unhinges Morford and others like him isn't a particular smell, it's the marketing of smell. Consumerism, mass consumption, and the excesses of the free market as embodied by a scent-delivery contraption really put his nose out of joint.

The psychoanalysts G. G. Wayne and A. A. Clinco offered a related criticism in 1959: "What was once a vital instrument for survival—directing and warning primitive man—has now deteriorated to an instrument for irrelevant and obtuse titillation through the double-jointed vocabulary of advertising." Emma Cook makes a similar claim: "Until recently, appealing to our sense of smell was relatively virgin territory for marketeers and manufacturers." (Cook missed the fact that her countryman Eugene Rimmel was marketing up a scented storm in the 1860s.) Common to all these critics is the notion that things were better in the good old days. They long for the unscented state of nature that existed before air fresheners, television, and perfume. Their olfactory Eden ended the moment one cavewoman asked another, "What are you wearing?" and traded a mastodon steak for a handful of aromatic resin. The fact is that millions of people enjoy giving their homes a pleasant scent, and, as in other areas of everyday life, they are willing to pay for convenience and a modest amount of fantasy.

I had a close encounter with anticapitalist scent-bashing a few years ago, when I was among a group of experts invited by the Smithsonian Institution in Washington to help the National Museum of Natural History plan a large traveling exhibit on the science and history of smell. Along with curators, exhibit designers, and high-ranking staff members, we spent the day in the museum's dark-paneled boardroom that looks out on Constitution Avenue and the IRS building. It was a typical institutional brainstorming session, with lots of cringe-inducing "exercises" meant to sharpen our creativity. One of these involved free association with pictures clipped from magazines. We took turns arranging them in domino fashion on the floor and afterward tried to interpret the pattern. The group decided the pictures fell into two categories: "human" and "environment." (I was puzzled; aren't humans part of the environment?) Then a senior curator reached down and removed an Estée Lauder soap ad from the arrangement; she felt it didn't belong to either category. I grew more puzzled.

For the next exercise, we broke into working groups. The soap-snatcher and I were assigned to the same group. Our task was to think of exhibit topics that would interest teenage visitors. With no prompting, she launched into a heated speech: the exhibit should make teens aware of how companies use smell to influence them. Others in the group gently challenged her, but she wouldn't relent. Her mission was to alert teens to the sinister corporate conspiracy behind fragrance advertising. I pointed out that subliminal advertising was largely a crock, but still she wouldn't let go. She was determined to stop America's youth from being turned into scent-controlled mall zombies. Finally, I reminded her that the Smithsonian was planning to fund the show with donations from corporate sponsors, and that these folks might be reluctant to fork over three million dollars for the privilege of having their business smeared.

The Smithsonian never did get around to doing a smell exhibition.

FOR EVERY ANTAGONIST of scent marketing there are a dozen crazily optimistic Martin Lindstroms preaching the benefits of sensory branding and experimenting with new ways of appealing to consumers through the nose. It's true that scent marketing has been promoted many times by futurologists of the past—it's a field whose promise has yet to be fulfilled. But the same can be said of Internet advertising or other new frontiers. The strategies of scent marketing are still evolving, but its technology has matured rapidly. All sorts of scent-delivery devices are available today, ranging from industrial-scale diffusers that cover an entire Wal-Mart to point-of-sale displays that blow a scented kiss at individual customers. There are passively activated devices that spritz as you walk past, and interactive kiosks that immerse you in a multisensory audio-visual-olfactory experience. Marketers will soon learn the best ways to put this hardware to use.

There is another reason to believe the field has a bright future. We are now raising a generation of scent-centric young consumers. Unilever's Axe body spray is a major hit: walk past any high school and smell for yourself. Aromatherapy has evolved from a quasi-clinical folk practice to mainstream product positioning; no college dorm room is complete without an array of scented candles. Students use them for studying, for chilling, and for, well, you know. So scent-aware is this generation that Procter & Gamble's Febreze odor eliminator is equally popular—and often seen in the same dorm rooms. These are the consumers who will put scent marketing on the map.

CHAPTER 10

Recovered Memories

To the boy Henry Adams, summer was drunken. Among
senses, smell was the strongest—smell of hot pine-woods
and sweet-fern in the scorching summer noon; of new-
mown hay; of ploughed earth; of box hedges; of peaches,
lilacs, syringas; of stables, barns, cow-yards; of salt water
and low tide on the marshes; nothing came amiss.

—HENRY ADAMS, *The Education of Henry Adams:*
An Autobiography (1918)

WHO HAS NOT ENCOUNTERED A LONG-FORGOTTEN
odor that brings to mind suddenly, and with great clarity,
a moment from the past? It leaves one marveling at the
potency—and persistence—of smell memory. It's an experience
people are eager to share with me. A compilation of their stories
would make a great autobiography of the nation's collective nose.
The American essayist Ellen Burns Sherman had a similar idea:
"Were they all collected in a volume, what a golden treasury of
poetry and romance would be the thousand records, grave, sweet and
tender, which are evoked from every one's past by the swift coupling
line of olfactory association."

Conventional wisdom credits the French novelist Marcel Proust
with the first literary description of the link between smell and

memory. His well-known account appears in the opening pages of his multivolume novel *Remembrance of Things Past* (1913), when the scent of a madeleine dipped in tea awakens childhood memories for the narrator, Marcel. A madeleine is a scallop-shaped sponge cookie—a bite-sized Hostess Twinkie without the filling, and without much flavor. That Proust constructed a 3,000-page story around it is, by itself, one measure of his literary genius.

The madeleine episode has become a cultural touchstone for the smell-memory experience. The poet Diane Ackerman calls him "that voluptuary of smell" and a "great blazer of scent trails through the wilderness of luxury and memory." The psychologist Rachel Herz claims, "Proust may have been prescient in noting the relationship between olfaction and the phenomenological experience of reliving emotions of the past." The science essayist Jonah Lehrer believes Proust revealed "basic truths" about memory, specifically that it "has a unique relationship" with the sense of smell. Lehrer credits the novelist with arriving at these truths before scientists did; in fact, he says "Proust was a neuroscientist."

Psychologists have made Proust their mascot for smell memory. Psychology journals are full of brand-conscious titles like "Proust nose best: Odors are better cues of autobiographical memory" and "Odors and the remembrance of things past." One has to admire how thoroughly Proust cornered this market—no other novelist has a branch of science named after him. Skepticism being one of the chief values of science, this sort of cheerleading makes one wonder whether Proust's insights justify the hero worship. Was he really the first writer to note a link between smell and memory? Did he really foreshadow modern neuroscience? To find the answers, we need to look more closely at Proust's original account.

THE ICONIC MADELEINE passage was published in 1913 in *Swann's Way*, the first installment of *Remembrance of Things Past*. A grown

Marcel is served tea and a madeleine by his mother. When he lifts a spoonful of tea and cookie to his lips, he shudders and feels an "all-powerful joy": "An exquisite pleasure had invaded my senses, something isolated, detached, with no suggestion of its origin." Marcel is overwhelmed by a nonspecific sense of familiarity. The smell and taste of the madeleine have something to do with it, but are not enough to evoke a specific memory. Marcel struggles to pinpoint the source of his déjà-smell. He tastes the madeleine again, plugs his ears, and tries to relive the initial experience. Finally, after two pages of strenuous effort, it comes back to him. When he was a child, his aunt Léonie would give him, on Sunday mornings, a piece of madeleine dipped in her tea.

Proust's struggle with the soggy madeleine is distinctly *not* the way most people experience odor-evoked memory. For most of us, these recollections spring to mind easily. We experience Sherman's "swift coupling line of olfactory association," not a prolonged, constipated mental effort. The smell scholar Dan McKenzie captures the feeling of effortlessness: "This strange revival of bygone days by olfaction is . . . automatic. It is most clearly and completely to be realised when the inciting odour comes upon us unawares, and then as in a dream the whole of the long-forgotten incident is displayed, even although it may have been an incident in which the odour itself was not specially obtrusive."

Here is another remarkable thing about the madeleine episode: it is utterly devoid of sensory description. Across four pages of text, Proust, that "voluptuary of smell," provides not a single adjective of smell or taste, not a word about the flavor of the cookie or tea. This is hard to square with his reputation as the sensual bard of scent. Outside of psychology, in fact, the experts are more impressed with his visual imagery. The literary scholar Roger Shattuck, for example, thinks that Proust's dominant mode of description is visual. Shattuck took a close look at the eruptions of involuntary memory that Proust called reminiscences or resurrections (*moments bienheureux*).

Of eleven examples in the entire novel, only two are triggered by smell, the madeleine incident being one of them.

Victor Graham is another scholar who finds that Proust's sensory imagery is largely visual. Graham indexed all 4,578 sensory impressions in the novel and found that 62 percent were visual. Smell and taste together accounted for less than 1 percent. This seems shockingly low, but it is on a par with other writers. In 1898 an obsessive psychologist named Mary Grace Caldwell tabulated every sensory adjective in the poetry of Shelley and Keats. She found that visual descriptors predominated: 79.9 percent for Shelley, 73.7 percent for Keats. Smell barely registered: 1.8 percent for Shelley and 2.7 percent for Keats.

Despite his reputation, Diane Ackerman's "great blazer of scent trails" was no more nasal than the next guy; nor did he write about smells very well. As Graham pointed out, Proust liked involuntary memories because they called forth "a flood of visual images" and emotions, but the flood contained very little aroma. Proust's trademark as a writer was to observe the recovery of a memory in excruciating detail, though after 3,000 pages it's not clear whether Marcel even liked the taste of madeleines. He was more interested in the process of introspection than in the smells it dredged up.

If Proust's reputation for psychological accuracy is questionable, what about the common assumption that he was the first author to recognize a powerful link between scent and memory? The record is clear, and it does not favor Proust. In American literature the memory-evoking power of smell was a commonplace observation long before *Swann's Way*. Sixty-nine years earlier, for example, Edgar Allan Poe wrote, "I believe that odors have an altogether idiosyncratic force, in affecting us through association; a force differing *essentially* from that of objects addressing the touch, the taste, the sight, or the hearing."

In 1851 Nathaniel Hawthorne expressed the same idea in *The House of the Seven Gables*: " 'Ah!—let me see!—let me hold it!' cried the guest, eagerly seizing the flower, which by the spell peculiar to remembered odors, brought innumerable associations along with the fragrance it exhaled."

In 1858 Oliver Wendell Holmes called attention to odor memory in his collection of essays *The Autocrat of the Breakfast Table*: "Memory, imagination, old sentiments and associations, are more readily reached through the sense of SMELL than by almost any other channel." Holmes illustrated his observation with an example from his own life. It's a sensory rhapsody of childhood in Cambridge, Massachusetts, sometime before 1825:

> Ah me! what strains and strophes of unwritten verse pulsate through my soul when I open a certain closet in the ancient house where I was born! On its shelves used to lie bundles of sweet-marjoram and pennyroyal and lavender and mint and catnip; there apples were stored until their seeds should grow black, which happy period there were sharp little milk-teeth always ready to anticipate; there peaches lay in the dark, thinking of the sunshine they had lost, until, like the hearts of saints that dream of heaven in their sorrow, they grew fragrant as the breath of angels. The odorous echo of a score of dead summers lingers yet in those dim recesses.

Holmes was a practicing physician as well as a writer. From his medical training he was well aware of the neuroanatomical basis of odor perception, and he had the Autocrat himself discuss it:

> There may be a physical reason for the strange connection between the sense of smell and the mind. The olfactory nerve— so my friend, the Professor, tells me—is the only one directly connected with the hemispheres of the brain, the parts in which, as we have every reason to believe, the intellectual processes are performed. To speak more truly, the olfactory "nerve" is not a

nerve at all, he says, but a part of the brain, in intimate connection with its anterior lobes.

The Professor contrasts this with the wiring of the gustatory system to explain why smell has a powerful link to memory but taste does not. Holmes's understanding of brain function is correct and modern—and it was written fifty-five years before *Swann's Way*.

While Proust was working on his novel, other writers were exploring the smell-memory connection. In 1903 the American physician Louise Fiske Bryson wrote, in *Harper's Bazaar,* "An odor, a perfume, will serve to recall bright scenes of other days with a vividness that is almost a miracle." In 1908 *The Spectator* published the essay "Scent and Memory," which used the image of a magic-carpet ride to describe how a sudden scent makes "miles of distance and decades of years vanish." Five years later Proust likened smell memory to being magically transported by a genie from the Arabian Nights.

Ellen Burns Sherman's thoroughly psychological account of odor memory was published in 1910, three years before *Swann's Way*. She described how an emotional moment woven into a man's memory along with the scent of his lover's perfume is brought to mind decades later when he catches "an infinitesimal whiff of the fragrance." Sherman says the former scene appears instantaneously, as if with "the turn of an electrical switch." In 1913 the American popular science writer Ellwood Hendrick, writing in *The Atlantic Monthly,* said, "These flashes of memory aided by smell are wonderful. Through smell we achieve a sense of the past."

Clearly, the subject of scent and recovered memory was very much in the air during the first years of the twentieth century. Proust shared this fascination and gave it his characteristic introspective literary treatment. For anyone not wearing Proust goggles, however, he was obviously not the first author to anticipate the discoveries of modern neuroscience.

How SECURE IS Proust's reputation as an olfactory innovator, if all these Yankees were saying the same thing years earlier? Perhaps he was the first French author to capture the phenomenon? Ah, *mais non*. The French author Louis-François Ramond de Carbonnières (1755–1827) was well known in Proust's day. In his most famous work, *Travels in the Pyrenees*, he described his descent from a mountaintop glacier on the border between France and Spain. He became intoxicated with the rustic smells of newly mown hay and flowering linden trees. As night fell, he tried to account for "the sweet and voluptuous sensation" that came upon him with such involuntary insistence. "There is something mysterious in odors which powerfully awaken the remembrance of the past. . . . The odor of a violet restores to the soul enjoyments of many springtimes." This has a Proustian ring to it, and for good reason. As the historian and critic Charles Rosen points out, "The coincidence is not fortuitous: Proust knew this page of Ramond." It was anthologized in French high school textbooks until very late in the nineteenth century.

Contemporary French psychology is another possible source of Proustian insight. Introspection was the research technique of choice—studies were done with one or two subjects trained to report their mental experience in precise detail. This emphasis on self-observed mental processing, of narrating one's inward gaze, is similar to Proust's "modernist" literary style. Théodule Ribot was the founder of modern scientific psychology in France; his 1896 book on the psychology of the emotions included a chapter on olfactory memory, which had been published earlier in the widely read *Revue Philosophique*. Ribot discussed such "Proustian" matters as odor memory, mental imagery for smell and taste, smell dreams, and smell hallucinations. The *Revue* was read not only by scientists but by the educated public, and Proust, who devoured periodicals, likely knew of it.

Between 1901 and 1903 the *Revue* published several articles on emotional memory. One, by a twenty-one-year-old French psychologist named Henri Piéron, contained this observation: "Sometimes, when passing through a certain place, while in a certain physical or mental state, I perceive a scent that, by itself, cannot be expressed or determined, that does not fit into the classification of odors; a composite, mixed scent that suddenly and violently plunges me in an indefinable, completely inexplicable but clearly felt and recognized emotional state." This sounds a lot like Proust's version of smell memory—all that's missing is the madeleine. (Piéron went on to coauthor a textbook and become *un grand frommage* in French psychology.)

Roger Shattuck identifies yet another French source of Proust's inspiration. In 1896 the philosopher Henri Bergson published *Matter and Memory*, a treatise on psychology that gained wide public attention. The nature of memory was at the core of Bergson's psychology, and he stressed in particular "pure or spontaneous memory," i.e., personal memories that survive in the unconscious for a long time before being recovered. The similarity to Proust's involuntary memory was obvious enough that Proust was asked about it in an interview in 1913. He denied being influenced by Bergson, a denial that Shattuck says "can only be termed ingenuous."

Marc Weiner, a professor of Germanic Studies at Indiana University, offers the sinister speculation that Proust lifted the tea-and-madeleine idea from Richard Wagner. When the composer was exiled from Germany for his political activities, he was unable to find any authentic zwieback biscuits. This led to a severe creative blockage while he was working on *Tristan und Isolde*. One day he received a shipment of real zwieback from Mathilde Wesendonk, his muse and platonic lover. In a letter, Wagner tells her (tongue-in-cheek) of the miraculous effects of her care package; how, when dipped in milk, the zwieback cured his writer's block and inspired him to move ahead with the opera. The Wagner-Wesendonk letters were widely read at the turn of the century; a French edition was published in 1905, eight years before *Swann's Way*. Weiner mischievously suggests that Proust's madeleine-soaking was inspired by Wagner's zwieback-dunking.

The Proust Boosters

Though Proust's notion of smell memory isn't very original, that hasn't stopped psychologists from adopting it with enthusiasm. The first researcher to charge forth under the banner of the soggy madeleine was Brown University's Trygg Engen. In a 1973 paper in the *Journal of Experimental Psychology*, he said, "The Proustian view is that odors are not forgotten to the same extent as are other perceptual events. Is there any factual validity for this claim of the artist?" Engen reported that the ability to recognize a set of memorized odors, though not high to begin with, did not drop off much over the course of several weeks. He concluded, "The Proustian insight is validated!" (His exclamation point, not mine.)

Engen's claim that odor memory doesn't decay was newsworthy. Mainstream memory theory in the 1970s was based almost exclusively on tests using words or pictures; memory for these stimuli faded according to well-known timetables. Yet from the beginning, smell psychologists assumed that odor memory was unique, a view steeped in conventional wisdom and garnished with anecdotes. Reviewing this period, Judith Annett notes that "negative experimental results were often taken to support the 'Proustian' position." The Proustian consensus that emerged in the 1970s—that odor memory decayed slowly if at all, and was unchanged by later experience—turns out to be wrong on both counts.

Engen's notion of indelible olfactory memory began to unravel in the 1980s. Heidi Walk and Elizabeth Johns, of Queen's University in Ontario, observed classic interference effects—smelling a second odor soon after the first makes the first one harder to remember. Others found that rates of forgetting were the same for odors as for sights and sounds. Odor memory appeared "to be governed by the same principles as remembering stimuli in other modalities." Such principles include interference effects and so-called rehearsal effects (an improvement in memory brought about by verbally describing the to-be-remembered odor). Most subsequent research, as the psychologist Theresa White has pointed out, shows that olfactory memory obeys the same rules as memory in the other senses: it erodes with time and is muddied by subsequent experience. The purity and infallibility of smell memory—an insight central to Proust's literary conceit—doesn't hold up to scientific scrutiny.

HAVING ROLLED SNAKE-EYES on their first Proustian bet, psychologists pushed their chips onto another. They proposed that personal memories elicited by odor were older and more emotion-laden than those sparked by words or pictures. The new experimental strategy was to give someone a smell, ask him to come up with a personal memory about it, and then rate that memory for age and strength of feeling.

Chief among the second generation of Proust Boosters was Rachel Herz, another Brown University psychologist who in one study asserted that she had produced "the first unequivocal demonstration that naturalistic memories evoked by odors are more emotional than memories evoked by other cues." Her bold claim deserves a close look. Herz asked people to recall a personal memory after she gave them an odor or a picture. People then rated their memories for emotionality. Picture-prompted memories had lower emotionality scores than odor-prompted ones, giving rise to Herz's claim. What she glosses over is the fact that both types of memory scored below the midpoint of the rating scale. In other words, visual memory and odor memory were both on the unemotional side of the scale. The odor-cued memories were simply *less unemotional*.

The Swedish psychologists Johan Willander and Maria Larsson have failed to confirm Herz's results. They cued autobiographical memories with odors, words, and pictures, and found that picture-evoked memories were the most emotional and odor-evoked ones were the least emotional. Willander and Larsson write that "we did not find support for the notion that olfactory-evoked memory representations should be more emotional than memories evoked by other sensory cues." It now looks as though the modified Proustian hypothesis—that odor memory, while not indelible, is more emotional—doesn't hold up too well either.

By 2000, the third generation of Proust Boosters arrived and wasted little time before turning on their predecessors. The British psychologists Simon Chu and John Downes criticized previous studies for being insufficiently Proustian. (They pointed out, for example, that the memories examined in some experiments were not truly autobiographical.) Chu and Downes contrasted those failed attempts with their own research agenda, which, in their modest view, captured the true spirit of Proust. Their goal was nothing less than "translating the essence of Proust's anecdotal literary descriptions into testable scientific hypotheses using the language of contemporary cognitive psychology." (This is a patently ridiculous thing for scientists to do. How can a work of fiction, no matter how well written, become the truth standard for scientific research? What's next? Will sex researchers lift hypotheses from Danielle Steel? Will Stephen King inspire psychiatric theories of fear?)

From out of left field came a quick challenge to Chu and Downes. J. Stephan Jellinek is a German psychologist who has worked as a perfumer and fragrance marketer. Not being an academic, he had the temerity to ask whether lab studies that relied on contrived and twice-prompted memories could capture the Proustian experience in any meaningful way. From a close reading of the madeleine episode, he extracts nine specific and testable characteristics of that experience. (Most have to do with the difficulty in identifying the emotion, tying it to an odor, and connecting the odor to an event in the past.) According to Jellinek, the experiments of Chu and Downes address only three of the key characteristics. Does measuring emotional response on a seven-point rating scale, he asks, truly capture the ecstatic experience described by Proust?

Determined to prove that odor memory is distinctive in some way, the latest Booster studies now claim that odors evoke older autobiographical memories than do words or pictures. This is an intriguing but ultimately trivial proposition. Whether this claim— the latest in a series of special pleadings—holds up is almost beside the point. Whether a lab experiment has captured the essence of Proust is certainly beside the point. The bigger question is why investigators decline to observe the natural history of smell for themselves, and prefer to base their research on a fictional episode. Three generations of psychologists have done so, and in each case they got lost in the woods. In the 1970s and 1980s the Proust Boosters grossly overestimated the permanence of odor memory. In the 1990s they overstated its emotional content. In the new century they overplayed how well lab studies could mimic an episode of fiction. Perhaps it's time for them to set aside the soggy Twinkie.

MEANWHILE, OUT IN the real world, a lot of people think that odor memory is special. A Norwegian survey recently compared popular beliefs with scientific findings regarding memory. Among the general population, 36 percent believed—incorrectly—that smells were remembered better than sights or sounds. This may reflect the fact that there is something unsatisfying about the current scientific view. If odor memory is like other forms of memory, why does it feel so magical when a sniff triggers a twinge of remembrance? A lot of it has to do with surprise. You weren't trying to remember the paints, oils, and solvents in Grandpa's workshop—the memory popped up, unasked for, when you walked through a random odor plume. Even more surprising: you never made a deliberate effort to memorize those smells when you were seven years old. If you had, the recollection would be no surprise. In grade school you memorized the state capitals; to recall one years later doesn't feel magical. Because odor memories accumulate automatically, outside of awareness, they cover their own tracks. We don't remember remembering them. The sense of wonder that comes with the experience is, like all magic, an illusion based on misdirection. Like a nightclub mentalist, the mind presents us with a memory it picked from our pocket when we weren't looking.

Henry Adams: The American Alternative

Psychology's preoccupation with Proust has led to a narrow emphasis on involuntary memory and a neglect of the far more common features of the mental smellscape. These include how and why we willingly commit some smells to memory and not others; how and how well we retrieve them; and how fully we are able to reexperience them. These questions are a promising starting point for a fresh exploration of olfactory memory.

If Marcel Proust is the poster boy for private, involuntary odor memory, this new alternative view will need its own mascot. I propose the American author Henry Adams, who conveyed in one sentence the actual sensations of a childhood smellscape. In his autobiography, written in the third person, he gave us a litany of scents from a boyhood in the days before the Civil War. As we return with him and stand beside the barefoot kid of summer, we feel his love of the outdoors: not for him the scent of inky copybooks or Mama's perfume, lavender sachets in the linen closet or bread in the oven.

Henry Adams gives us a small sample of a true olfactory

memoir—it puts you behind another person's nose in another time and place. In his honor, I call it Adams-style odor memory. To my way of thinking, Adams-style memory beats Proustian memory because it deals with smells that are deliberately sniffed and voluntarily recalled. These are not the buried land mines of Proustian memory; Henry Adams describes a smellscape that was familiar to his entire generation, and his memory of it is open to the public. Proustian memory inhabits a private, interior place, and is open by invitation only. For Proust, smell was a tool, a reflex hammer he used to probe his own mind. For the young Henry Adams, smell was the whole world; for the old Henry Adams, it was an open gateway to the past. Breathe deep: it's summer, the sun is hot, and the tide is low.

Adams-style odor memory is popular with American writers. A fine example is found in the opening lines of *Lake Wobegon Days,* where Garrison Keillor conjures up the fictional town of Lake Wobegon, Minnesota:

> Along the ragged dirt path between the asphalt and the grass, a child slowly walks to Ralph's Grocery, kicking an asphalt chunk ahead of him. It is a chunk that after four blocks he is now mesmerized by, to which he is completely dedicated. At Bunsen Motors, the sidewalk begins. A breeze off the lake brings a sweet air of mud and rotting wood, a slight fishy smell, and picks up the sweetness of old grease, a sharp whiff of gasoline, fresh tires, spring dust, and, from across the street, the faint essence of tuna hotdish at the Chatterbox Cafe.

You don't have to be a Norwegian bachelor farmer to appreciate this. Anybody can inhale the scene and experience Lake Wobegon.

Adams-style memory has a big scope: it's about extended episodes, not single events, entire smellscapes rather than isolated odors. Adams-style memory edits an entire season down to an aromatic highlight reel that can replayed at will. Dozens of Saturday afternoons with Grandpa at his workbench are distilled into a few key molecules.

By preserving familiar scenes, Henry Adams left us a time capsule of a lifestyle that has nearly vanished. For most of our history, most Americans lived and worked on farms; agriculture was our common smellscape. Haydn Pearson was born in 1901 and grew up on a small family farm in Hancock, New Hampshire. In a memoir, he recalls the ambience: "When I was a boy, one of my favorite spots was the livery stable. When I walked into Woodward's Livery behind the Forest House Hotel, I was met by a pungent heady fragrance compounded of hay, leather, grain, harnesses, stained and splintered floor planks, and manure." The interior of the livery office had its own character, "the fragrance of felt leggings, rubber arctics, sheep-lined coats, and the sawdust box for tobacco juice blended very pleasantly with the over-all aroma of the establishment."

His family stored root vegetables and preserved foods in the farmhouse cellar, which acquired its own atmosphere: "a heavy damp pungent smell compounded of moist soil, potatoes, apples, carrots, turnips, salt pork, cold crackling brine, and the old floor boards. Probably there were some rotten potatoes and possibly a decayed cabbage or two, and if there is any farm-cellar fragrance equal to the combination of decayed potatoes and decomposed cabbages, I have yet to smell it."

For Ben Logan, born in 1920 and raised on a small farm near the Kickapoo River in southwestern Wisconsin, haying time was aromatic: "A time like that comes back now sharp and real with all its smells of dust, horse sweat, man sweat, Lyle's oozing pipe. There is the dry whirring of grasshoppers, steel wagon wheels ringing on the hard ground, the creak of the hay rope. There is the tepid smell of water as we drink from a bucket that has a taste of leftover lemonade. Above all is the sweet smell of curing hay."

Proustian memory is involuntary; we have no control over its recording or its recall. Because it is recoverable on demand, Adams-style odor memory is a more useful storage medium—it embodies our common past and gives us a way to preserve it. Some people improvise their own olfactory scrapbook. An attorney who was in his thirties at the time once described his method for inducing scent-fueled visions of the past:

> I grew up on the Nevada desert in a small mining town. Since my seventeenth year my residence has been in California in the San Francisco bay area but I never have and never will learn to be happy in the fog and rain and dampness. I have a perpetual nostalgia for the sun, warmth, clear, clean air, the peculiar lemon desert fragrances and the great panoramic vistas and strong colors. I have spent part of several summers in the Tahoe district and each time have brought home a good bunch of sage brush which I keep in a receptacle and not infrequently smell. When I do, visual and emotional sensations arise within me in considerable clarity of the desert scene. A slight sniff doubles and redoubles that tranquil nostalgia.

The scientific study of smell memory is currently in flux. After a long and fruitless detour spent quantifying a literary fiction, the field is abandoning the idea that smell is unique among the senses. Just as the larger field of memory research has retreated from the notion of indelible flashbulb memory and questioned the veracity of eyewitness testimony, smell experts are recognizing that memory for odor is like memory for anything else—subject to fading, distortion, and misinterpretation. With this realization, we give up some long-held ideas, but throw open the windows for a breath of fresh air.

The Smell Museum

My collection of semi-used perfumes is very big by now, although I didn't start wearing lots of them until the early '60s. Before that the smells in my life were all just whatever happened to hit my nose by chance. But then I realized I had to have a kind of smell museum so certain smells wouldn't get lost forever.

—ANDY WARHOL

ANDY WARHOL MAY HAVE SAVED MODERN CULTURE without even realizing it.

Memories fade and get harder to find amid the mental clutter of a busy life. For a given smell, the odds that it will produce a riveting flashback shrink with each resniffing. That special scent becomes less special, its links to the past grow steadily weaker. Warhol's solution was ingenious: he would wear a cologne until it built up strong emotional connections, then retire it to his personal smell museum. Once out of active rotation, the cologne's memories were locked in, never to be confused with others. The Warhol wear-and-retire method was unusual but effective. By not switching back and forth between scents, he avoided the loss of memorability that psychologists call interference.

It's easy to reach into the past when the missing link sits on a shelf, clearly labeled. But even a cologne collection has its limits—brands don't live forever. Commercial death occurs when the last bottle comes off the production line, and psychosensory rigor mortis sets in with the last spray from the last bottle. An extinct fragrance triggers no memories. To preserve links to the past, we must preserve the juice itself. How will we know what we're missing when it's not there to smell?

The James Joyce scholar Bernard Benstock concludes that the juice doesn't matter as long as we have literature: "[E]ach work of fiction is posterity-proof. No captured smell specified in *Ulysses* is ever lost in the rereading or fails to register its full pungency for every new reader." Why is Professor Benstock so sure that every reader gets a noseful from the novel? This seems like wishful thinking. A reader may be able to reimagine a familiar smell, but for one he doesn't know, he's left to guess. To reexperience the smells of times gone by, one needs the *actual stuff*; without it, written references and therefore literature eventually lose their power.

"Cannery Row in Monterey in California is a poem, a stink, a grating noise, a quality of light, a tone, a habit, a nostalgia, a dream." The opening line of John Steinbeck's 1945 novel acknowledged the reek of the fish-processing plants on Cannery Row, but by the 1950s, overfishing had flattened the local sardine population and taken the factories down with it. When he returned to Monterey in 1960, Steinbeck climbed up Fremont Peak for a last panoramic look at the land of his youth. The canneries had disappeared and so had their "sickening stench"; all that was left was the smell of wild oats on the dry brown hills. It brought to his mind Tom Wolfe's phrase: you can't go home again. Steinbeck had immortalized the smell of Cannery Row on the printed page, but he could no longer inhale the thing itself—and neither could his readers.

When an entire smellscape fades away, especially one familiar to many people, our culture suffers a loss. Take the case of the local tavern. The journalist and pundit H. L. Mencken grew up in Baltimore and accompanied his father—a cigar manufacturer—to the saloons where he sold his product: "In the days before Prohibition, which were also the days before air-cooling, I doted on the cool, refreshing scent of a good saloon on a hot Summer day, with its delicate over-tones of mint, cloves, hops, Angostura bitters, horse-radish, *Blut-wurst* and *Kartoffelsalat*. It was always somewhat dark therein, and there was an icy and comforting sweat upon the glasses."

Mencken couldn't relive his memories in today's gleaming, art-fully designed modern brew-pub, but he might feel at home in a place like McSorley's Tavern on Manhattan's Lower East Side, which has been serving ale in an atmosphere little changed since it opened in 1854. Patrons find something soothing in its quiet, almost gloomy interior. As one regular described it, in 1943, "there is a thick, musty smell that acts as a balm to jerky nerves; it is really a rich compound of the smells of pine sawdust, tap drippings, pipe tobacco, coal smoke, and onions. A Bellevue intern once said that for many mental states the smell in McSorley's would be a lot more beneficial than psychoanalysis or sedative pills or prayer." Coal-burning furnaces disappeared decades ago, and in 2003 the city's mayor banished the sweet, warm notes of tobacco, yet McSorley's retains its distinctive aroma: a dark, hoppy yeastiness livened by the sawdust on the floor. TGI Friday's it's not. McSorley's is the Kong Island of taverns, a place where prehistory lives on—for now.

High on the list of endangered smellscapes is the heartwarming aroma of Grandma's kitchen. Fewer families eat dinner at home, and when they do, they don't cook: they microwave frozen food, which doesn't pack the same emotional punch. The aroma of a tomato sauce simmering all day? Fuhgetaboutit. Chicken roasting in the oven? No one has the time. Apple pie? Pick it up at the A&P. Coffee aroma? Kiss it good-bye: half of Americans in their thirties get their hot java at a store; the proportion is even higher for those under thirty. Home-brewed coffee will soon be a game for the elderly.

The extinction of familiar smells leaves the fabric of our culture looking rather moth-eaten. It even affects movie watching. Take the scene in *Fast Times at Ridgemont High* where a classroom full of students plunge their faces into quiz papers fresh off the ditto machine. The visual joke is lost on anyone born after 1982. The Wite-Out-sniffing school secretary in *Ferris Bueller's Day Off* will be equally incomprehensible: Correction fluid died with the typewriter.

When most Americans lived on farms, cow manure smelled of income and family security. In rural areas today, newly arrived suburbanites feel differently; they consider dairy farms a public nuisance, and object to the spreading of manure on fields. To defend farming as a way of life, the Planning Commission in Ottawa County, Michigan, put a manure-scented scratch-and-sniff panel in an explanatory brochure for people moving into the area. Lebanon County, Pennsylvania, followed suit with its own smellustrated pamphlet.

IT IS THE natural order of things for smell preferences to change from generation to generation. Back in 1931, a survey ranked the popularity of fifty-five commonplace odors. The results were not surprising: pine, lilac, rose, and violet were at the top, garlic and perspiration at the bottom. It is odd to look back at some of the other smells included in the survey: witch hazel, sarsaparilla, lard, and turpentine. These were commonplace seventy-seven years ago, but today they seem exotic. When did the last drop of sarsaparilla evaporate from the national smellscape? Did it outlive witch hazel? It would be enlightening to track changes in odor perception and public opinion over the long term. What we need is a Scent Census.

The architect Rem Koolhaas knows how rapidly a smellscape can vanish. "I turned eight in the harbour of Singapore. We did not go ashore, but I remember the smell—sweetness and rot, both overwhelming. Last year I went again. The smell was gone. In fact, Singapore was gone, scrapped, rebuilt. There was a completely new town there."

In the Northeastern United States the smell of burning leaves was once emblematic of autumn. Everyone understood Booth Tarkington's allusion to it in *The Magnificent Ambersons*: "When Lucy came home the autumn was far enough advanced to smell of burning leaves, and for the annual editorials, in the papers, on the purple haze, the golden branches, the ruddy fruit, and the pleasure of long tramps in the brown forest." The lazy plume of gray smoke from a smoldering leaf pile accompanied the mood of a declining season, a time of endings, sadness, and reflection. Edgar Lee Masters used it to depict an old man's melancholy: "Now, the smell of the autumn smoke, / And the dropping acorns, / And the echoes about the vales / Bring dreams of life."

By now, several generations of children have grown up without burning leaves. The scientist and physician Lewis Thomas thinks this is a shame: "[W]e should be hanging on to some of the great smells left to us, and I would vote for the preservation of leaf bonfires, by law if necessary." For Thomas, playing by a curbside bonfire was fun and risky—the perfect childhood activity. "It was a mistake to change this, smoke or no smoke, carbon dioxide and the greenhouse effect or whatever; it was a loss to give up the burning of autumn leaves." Environmentalist sensibilities be damned; Thomas wanted to empty the leaf bags and toss a lit match. His nostalgic fantasy is unlikely to come true; few will ever know the acrid smoke and quiet crackle of burning leaves. The old incense of suburban lawn worship has been replaced by the new roar of leaf blowers and the fumes of half-burned gasoline.

A Blast from the Past

The need to preserve today's smells might not seem urgent—after all, we can always use technology to recover the past. The trouble is that it takes an extraordinary effort to re-create an extinct smell.

Take, for example, a 1984 study in which researchers tried to revive food aromas in order to study the composition of prehistoric diets. The smells they were after were locked into a fossilized human turd (politely known as a coprolite). The specimen in question was deposited on a cave floor in Utah about 6,400 years ago. Perfectly preserved by the desert climate, it presented the scientists with a challenge: there was no established protocol for resuscitating ancient poop. Accordingly, the research team spent a month inventing and perfecting their own technique. The first task was to produce a set of reference stool samples for training purposes. They did this by feeding a series of controlled meals (high fiber, mixed fruit and vegetable, peach only, etc.) to a selfless volunteer who saved the resulting output. His contributions were freeze-dried to create pseudo-fossils for pilot testing. To make the practice samples sniffable, they were soaked in a solution of trisodium phosphate until they released enough aroma for analysis. (Note to students planning science fair projects—this step takes a few days.) An experienced sniffer took notes as the volatiles exited the gas chromatograph. Out came a rainbow of aromas: bread, corn, peanut, beer, peach, popcorn, onion, licorice, cauliflower, and meat. The more things the volunteer ate, the more smells the team detected.

Having perfected their technique, the team was ready to analyze the turd of historical interest. They placed the ancient sample in the GC and waited for it to yield its secrets. One can imagine the tension in the lab as the instrument warmed up and the researchers hovered over the exhaust vent in anticipation. Would they get something, or was all their preparation in vain?

Within minutes secrets of the ancient bowel movement began to spill from the GC. The researchers got a noseful of the expected fecal notes, but along with them came an assortment of food aromas: green leaves, grass, and (weirdly) licorice. Next, they injected a sample from a more recent specimen, one found in Glen Canyon and dating from AD 1100 to 1300. From this one they smelled burned corn, meat, and, once again, licorice. The licorice smell was not an aberration; two plants native to the region smell of it, American licorice and sweet cicely, and both were eaten by Native Americans. Science has succeeded in turning the GC into a time portal. There are probably a lot of fossilized smells lying on museum shelves; which one will be reanimated next?

If You Build It . . .

With entire smellscapes going extinct, there is an urgent need for preservation. Can a scaled-up version of Warhol's personal smell museum solve our crisis of collective memory?

In Salinas, California, the National Steinbeck Center is attempting to preserve Steinbeck's marvelous fictional smellscapes. His inventory in *Cannery Row* of Doc's workroom in the Western Biological Laboratory, for example, is a sustained tracking shot for the reader's nose:

> Behind the office is a room where in aquaria are many living animals; there also are the microscopes and the slides and the drug cabinets, the cases of laboratory glass, the work benches and little motors, the chemicals. From this room come smells— formaline, and dry starfish, and sea water and menthol, carbolic acid and acetic acid, smell of brown wrapping paper and straw and rope, smell of chloroform and ether, smell of ozone from the motors, smell of fine steel and thin lubricant from the microscopes, smell of banana oil and rubber tubing, smell of drying wool socks and boots, sharp pungent smell of rattlesnakes, and musty frightening smell of rats. And through the back door comes the smell of kelp and barnacles when the tide is in.

On display at the Steinbeck Center are permanent interactive exhibits in which smells are matched to the books where they appear: horse stable for *The Red Pony,* mangrove flower for *The Log from the Sea of Cortez,* and so on. (The smells are released periodically from hidden aerosol cans operated by a timer.) Olfactory realism occasionally takes a back seat to ticket sales: the Cannery Row sardine smell proved too unpleasant for visitors, who complained that something in the museum was rotting. The smell of old dog that accompanies *Of Mice and Men* is also not popular, but the curators left it in.

Scented museum exhibits are not new; the Smithsonian snuck lavender into a display of gowns in the Hall of American Costume in 1967. Today the Tenement Museum on Manhattan's Lower East Side allegedly uses a scent generator to simulate the smell of a coal-burning stove in its restored 1878 tenement house. The idea is good—an overcrowded, unventilated apartment of that era would also have reeked of cooking food, BO, and chamber pots—but the execution is too faint to bring much life to the setting.

English museums are especially keen on smells; if you find yourself at a loss for entertainment in the coastal town of Grimsby, go to the National Fishing Heritage Centre and get a noseful of maritime history: seaweed, sea breeze, and dried codfish are among the offerings. Or head to York, where the Jorvik Centre uses smells to re-create life in a Viking village. At a maritime museum in Liverpool, the engine room of the restored pilot cutter *Edmund Gardner* is enlivened with the smells of diesel fuel and hot oil. In 2001 London's Natural History Museum pushed the curatorial envelope by creating foul dinosaur breath for a *T. rex* exhibit. At the last minute, however, the curators lost their nerve. They substituted a vague, nonthreatening boggy-swampy scent meant to evoke the Cretaceous environment of *T. rex.* If you close your eyes and breathe deeply, you might think you're standing in the New Jersey Meadowlands on a ripe day.

The uptick in scented exhibits is evidence of museums' eagerness to be less intimidating and more consumer-friendly, less like temples of culture and more like theme parks. Some aim for what the art critic Jim Drobnick calls "aromatopia": a total-immersion, firing-on-all-five-senses sort of experience for the paying public. In doing so, they go head-to-head with Las Vegas casinos and other venues, which, as we've seen, are heavily into sensory engineering.

PRESERVATION IS A priority for the fragrance industry, which bases its prestige on a long and continuous history of trend-setting creations, and which it expects its new recruits to learn. The world's most extensive perfume museum is the Osmothèque in Versailles, France, founded in 1990 as part of a training institute for fragrance, flavor, and cosmetics. There are more than 1,400 perfumes in the Osmothèque's collection, including 500 that are no longer manufactured. Despite having worked in the industry, I find it hard to get excited about visiting a perfume museum—how many little bottles can one stand to look at? (take one down, pass it around, 1,399 bottles of scent on the wall . . .) To some people, a vintage bottle of *Halston* is a fetish object; to me it has the emotional resonance of an empty Coors longneck. Still, 500 samples of extinct juice might be worth a stop, especially if they were presented in a compelling way, say a vertical sniffing ("From *Obsession* to *Euphoria*—A Calvin Klein Retrospective"), or a vintage sampling ("Backlash: Transparent Top Notes in the Post-*Giorgio* Years").

Touring a perfume museum would be a testosterone-draining ordeal for most men. As a gesture to them, if nothing else, I would suggest the museum include a hormone-stabilizing Hall of Technology. Displayed in a spotlight under a glass dome would be the first spray bottle invented by Jean Sales-Girons in 1859. Sale-Girons wasn't thinking about perfume: he wanted people to be able to inhale the allegedly therapeutic mineral waters of French spas. Later, his "vapeurisiteur" was adopted by physicians to spritz medicine into a patient's nose and throat. Other uses were found for the classic spray bottle with the rubber squeeze-bulb, and it soon became standard equipment for dentists, chemists, barbers, and other manly professionals. The atomizer underwent a dramatic sex change at the Paris Exposition Universelle of 1878. It was at this gigantic industrial trade show, according to the atomizer historian Tirza True Latimer, that it crossed over into consumer culture and became feminized. When Guerlain and other French perfume manufacturers at the show spritzed their latest creations onto the passing crowds, women immediately saw that misting was an excellent way to apply perfume—evenly and with no dripping onto clothes. By 1890 the atomizer was on ladies' dressing tables around the world, and remained so until the invention of the pump spray.

My ideal Hall of Technology would feature significant contributions to science and technology made by the perfume atomizer in masculine hands. Wilhelm Maybach of Germany, who was designing the first internal combustion engines in the late 1800s, needed to get gasoline into the cylinders in a way that would maximize its explosive force when ignited. His wife's perfume atomizer provided the inspiration for his invention of the carburetor. A few years later a University of Chicago graduate student named Harvey Fletcher was working with physicist Robert A. Millikan to measure the charge of the electron. They had been suspending particles of water vapor between two conducting plates, but the water was evaporating too quickly. They decided to try oil instead. Fletcher went to a jeweler's for watch oil and on impulse bought a perfume atomizer to create a fine vapor of oil droplets. The experiment worked and Millikan received the Nobel Prize for Physics in 1923.

The Hall of Technology would not be complete without an exhibit honoring Gale W. Matson. She was an organic chemist at the 3M company who was looking for new ways to make carbonless copy paper in the early 1960s. She ended up inventing scratch-and-sniff technology instead (both processes encapsulate tiny drops of liquid inside a burstable shell; ink in one, fragrance oil in the other.) Scratch-and-sniff was an immediate hit with children: *The Sweet*

Smell of Christmas (1970) is still in print, along with dozens of smelly baby books. Beginning with a June 1972 ad in *McCall's* for *Love's Lemon Fresh,* scratch-and-sniff was used for perfume ads until higher-fidelity methods came along.

Scratch-and-sniff excels at bringing out the grosser, masculine side of life. Larry Flynt, the fabulously vulgar publisher of *Hustler,* was an enthusiast. FIRST TIME EVER SCRATCH 'N' SNIFF CENTERFOLD, screamed the cover of his August 1977 issue. In smaller print at the bottom: "WARNING: To be smelled in the privacy of your home. Not to be smelled by minors." (The actual smells were G-rated: banana, rose, and baby powder.) The film director John Waters, of course, gave audiences scratch-and-sniff cards for *Polyester,* his 1981 homage to Smell-O-Vision. One of the first "adult" computer games—*Leather Goddesses of Phobos,* released in 1986—came with a seven-item scratch-and-sniff card and a big floppy disk for the Commodore computer. At various points the game instructed the player to sniff location-specific odors: mothballs in the closet, perfume in the harem, leather in the boudoir, etc. Probably the most testosterone-heavy scratch-and-sniff ad was run by the BEI Defense Systems Company in *Armed Forces Journal International.* With the tag line "The smell of victory," it touted the "battle-proven, state-of-the-art HYDRA 70 family of rockets" using the scent of burnt cordite.

I think the Hall of Olfactory Technology would be a major attraction, but something tells me it wouldn't find a happy home at the Osmothèque in Versailles. It might work better in Paris, Texas, where weekend crowds could roll in on Harley-Davidson Fat Boys, and reminisce about the sweet, long-lost smell of leaded gas.

THE ADDITION OF scent to museum exhibits raises a related question: where is the smell in traditional art? Olfactory art has never really taken off. Jim Drobnick suspects the concept is too novel to be accepted by museums and "serious" collectors. I disagree, given that the contemporary art establishment values the revolutionary, the challenging, and the transgressive above all else. Would not Andres Serrano's *Piss Christ*—a crucifix submerged in urine—have been even more transgressive if it smelled like stale pee?

Unfortunately, olfactory artwork teeters between banality and pretension. The former was on display in an installation by Alex Sandover in a New York gallery. A video screen showed a woman preparing dinner in a 1950s-style kitchen. As she worked, wall-mounted diffusers released the corresponding scent: sage, apple pie, etc. The see-it/smell-it conceit was literal-minded and certainly not very transgressive. (If his housewife had vomited on camera, with a scent-track to match, Sandover would have been an art-world hero.)

Sissel Tolaas, a Norwegian artist who lives in Berlin, gets closer to the mark. She collected underarm sweat from nine men who were in various states of fear and anxiety, chemically extracted their BO, had it microencapsulated, and then spread it onto large colored sheets. She mounts these enormous scratch-and-sniff panels on art gallery walls for visitors to sample. Her 2007 show is called "The Fear of Smell and the Smell of Fear." It sounds creepy and probably smells worse. Sissel Tolass could go far—she has a firm grasp of the transgressive.

Artists have a hard time incorporating smell into the traditionally visual arts. Scent hangs awkwardly in the air and strikes viewers as an afterthought. (Jackson Pollock—now with peppermint!) Visual artists may have a hard time putting into practice the smells they create in their imagination; one solution is to work with someone who has the know-how. In 2004 SoHo's Visionaire gallery paired celebrity photographers with perfumers and exhibited the results in a pitch-black gallery. Next to each backlit color photo was a nozzle and a button; pressing it released a puff of scent. In Karl Lagerfeld's photo, titled *Hunger,* a naked guy holds a round loaf of bread in front of his groin. The accompanying fragrance, by Sandrine

Mali, was rather mundane—neither yeasty nor beastly. Another entry, by celebrichef Jean-Georges Vongerichten and perfumer Loc Dong, called *Strange,* was a photo of a durian fruit split open to emphasize its resemblance to the female anatomy. The subtext was clear: "We double-dog dare you to sniff it." I did, and found that the highly abstract scent didn't carry through on the visual metaphor.

Olfactory art as performance art has the potential for embarrassing pretension. Mark Lewis's *Une Odeur de luxe* (1989), for example, sounds like a pretty good junior-high-school prank that was taken seriously by the grownups, including Jim Drobnick. Here's his account of it:

> Lewis's dialectical odours . . . attempt to expose and corrupt the ideology of sexual difference and what Lacan terms "urinary segregation." By atomizing women's perfume in the men's bathroom and men's cologne in the women's, Lewis interrogates the politics of identity construction and its performative maintenance. These transgendered diffusions of odour, rendering each space (and each person within it) olfactorily hermaphroditic, forces a confrontation with architecture's role in naturalizing sexual difference as an unproblematic binary opposition.

That's a tad more interpretation than *Une Odeur de luxe* can bear. I prefer to think of Mark Lewis as an art school Bart Simpson. Someone should make him write one hundred times on the blackboard, "I will not spray cologne in the girls bathroom."

Freak Show

While museum directors ponder whether olfactory art deserves gallery space, one smell has proven to be box-office gold: the stench of rotting flesh. This putrid but profitable aroma is emitted by a giant flower stalk, which people are willing to stand in line to see and smell up close. It's become the Lobster Boy of the olfactory sideshow.

The plant, *Amorphophallus titanum*, was discovered on the island of Sumatra in 1878. It spends most of its life underground as a large tuber weighing up to 170 pounds. Every two or three years it sends up a three-to-nine-foot-tall flower stalk called a spadex. Its Latin name means "huge shapeless penis," which gives you a fair idea of what it looks like. The fast-growing flower stalk lasts about three days and smells of dead meat; in nature the scent attracts blowflies, flesh flies, and carrion beetles. After these creatures pollinate the blossom, it stops producing scent and quickly shrivels.

A. titanum emerged from obscure, humid greenhouses to become a celebrity tuber. Dubbed the corpse-flower (allegedly the translation of its Sumatran name), it had limited exposure to the public before botanical gardens shared seedlings and made it into the porn star of the vegetable world. Its United States debut was at the New York Botanical Garden in 1937, but its big break came when a blossom at Kew Gardens in London drew 50,000 visitors in 1996. Four television crews reported on the specimen at the Atlanta Botanical Garden in 1998. Intense media coverage raised public expectations to unsustainable levels: "'It smells a little like dirty socks,' said John Allison of Marietta, who dropped by to see the bloom Monday with his wife, Joan. 'We expected rotting human flesh.'" Charming folks, the Allisons. I guess they left Pugsley and Wednesday at home with Uncle Fester.

Giant, evil-smelling penis-plants are performing everywhere. Dates and venues read like a rock tour: 1998 Atlanta and Miami, 1999 Sarasota and Los Angeles, 2001 Washington, D.C., and Madison, Wisconsin, with return shows in Miami and Atlanta. Media-savvy curators have turned up the hype. The Marie Selby Botanical Gardens in Sarasota posted blossom updates on its website. Not to be outdone, the University of Wisconsin put its bloom on a live webcam. As its popularity soared, *A. titanum* got an image makeover; the term "corpse flower" was quietly dropped and the plants were given personalities. In 2001 Miami named its blossom Mr. Stinky. UC Davis countered with Ted, followed by Tabatha in 2004. Cal State Fullerton trumped Tabatha with Tiffy. Tabatha drew only 4,000 live sniffing visitors, but pulled 52,000 hits on the website and 11,000 visits on the webcam. (This is puzzling: Why stare at Mr. Stinky online when you can't smell him?) Merchandising tie-ins are only a matter of time: "Hi, my name is Tiffy. You can watch me on my webcam, and buy my fragrance online."

Mapping the Smellscape

Rudyard Kipling memorialized the transporting power of scent in these widely quoted lines: "Smells are surer than sounds or sights / To make your heart-strings crack— / They start those awful voices o' nights / That whisper, 'Old man, come back!' " Where Proust was concerned with time, Kipling was concerned with space. His theme was homesickness; one smell encountered on two continents. Kipling wasn't being abstract—he had one particular smell in mind, and it shows up in the next, less quoted, stanza: "That must be why the big things pass / And the little things remain, / Like the smell of the wattle by Lichtenberg, / Riding in, in the rain." The smell of wattle, which appears in all five verses, is central to the poem. What, you might ask, is wattle and why did it have this profound effect?

"Lichtenberg" is told in the voice of an Australian trooper from New South Wales who is riding his horse in South Africa during the Boer War. Golden wattle is a plant—a small tree in the mimosa family. It is also the floral emblem of Australia. In the spring it develops a spectacular, golden-yellow flower head that throws off a heavy, floral scent with a honeylike sweetness. Kipling's inspiration was an incident that happened when he was in South Africa: "I saw this Australian trooper pull down a wattle-bough and smell it. So I rode alongside and asked him where he came from. He told me about himself, and added: 'I didn't know they had our wattle over here. It smells like home.' That gave me the general idea for the verses; then all I had to do was to sketch in the background in as few strokes as possible."

The power of smell to evoke a particular place gives the smell museum a unique opportunity for innovative exhibits. Perhaps something along the lines of a recent presentation by the designer Hilda Kozári and the perfumer Bertrand Duchafour. They linked scent and place in a 2006 artwork called *AIR—Urban Olfactory Installation*. Kozári suspended three translucent globes from the ceiling, each one large enough for a visitor to step into through a hole in the bottom. Around each globe's equator, a thin layer of spongy material was moistened with a city-scent composed by Duchafour. Monochrome video images were projected onto the sphere's surface. By standing inside, one could experience Budapest (Kozári's hometown), Helsinki (where she works), or Paris (just because).

Great balls of smell is a very cool concept. The light, leafy-green scent in the Helsinki ball was pleasingly matched by the green-tinted video. The smells of Budapest and Paris, however, were indistinct, and the three videos, shot from a moving car, made all the cities look the same—an endless loop of roads, bridges, and traffic. I entered the balls with high hopes, but left underwhelmed. I thought of Kipling's poem and yearned for a Lichtenberg experience; I wanted to smell wattle and watch it rain in Australia on one side of the globe, and in South Africa on the other.

IF WE'RE SERIOUS about preserving scents of place, it's not enough to capture random locations; we should survey an entire geographical area. I once accompanied a *New York Observer* reporter on a sniffing safari of Manhattan. It was midsummer and New York was ripe, but nailing down the actual source of the malodors wasn't easy. The air in an upscale sports club was a tad stale but not too objectionable. Our most noxious find was a puddle of rancid sidewalk water at University Place and Thirteenth Street. Something terrible had happened there, and the ghost of it lingered in the late afternoon. The *Observer* reporters conducted walking tours with other nose experts and published the story along with a whimsical odor map of the city.

The guided odor tour has become a features-section standby. For example, a *Washington Post* reporter rides along in a limo with a perfumer and a retired sanitation worker as they make a haphazard tour of New York. They hit the usual tourist sites with predictable results: rancid pork fat in the Meatpacking District, hot frying oil in a Chinatown kitchen, and intense horse manure near the carriages in Central Park. All the while the French-born perfumer plugs her line of neighborhoods-of-New-York-themed perfumes. (Fair enough—it was her limo and driver after all.)

The New York–based gossip blog Gawker took a refreshingly egalitarian approach to urban odor mapping. It invited readers to e-mail in odor reports for every train station and subway platform in the city. The general outcome was not in doubt. (Even Paris Hilton knows the score; in her memoir she writes, "Yes, I admit I've taken the subway in New York—and it smells. It literally smells like pee. Why can't they do anything about that?") Gawker compiled the vox populi into an interactive New York City Subway Smell Map. Mouse over a particular station, and colorful icons pop up to tell you which of ten malodor categories is found there. Waiting for the A-C-E train at Thirty-fourth Street and Eighth Avenue? Gawker icons indicate the presence of body odor, feces, urine, sewage, and vomit. Need more detail? Just double-click for reader comments: "Something dead and decaying . . . Old outhouse poop . . . Fresh poop . . . Sewer water . . . Urine post–asparagus buffet . . . Breath of a hungry old lady . . . Stinks like puke." According to the Subway

Smell Map, stations on the Upper East Side are exceptionally non-odorous. This may be true, or the result of sample bias—hipsters who read Gawker may never venture that far uptown.

The ultimate objective for nasal surveyors is a navigational chart of the entire American smellscape. Is such a thing possible? Helen Keller thought so: "I can easily distinguish Southern towns by the odours of fried chicken, grits, yams and cornbread, while in Northern towns the predominating odours are of doughnuts, corn beef hash, fishballs, and baked beans." American cities were so distinctive she had her own Olfactory Positioning System: "I used to be able to smell Duluth and St. Louis miles off by their breweries, and the fumes of the whiskey stills of Peoria, Illinois, used to wake me up at night if we passed within smelling distance of it."

Landmark smells, even those of home, are not always pretty. The writer Celeste Bowman describes her experience in Texas: "My eyes flew open as my nose was assaulted by the acrid odor of saltwater, decomposing fish and seashells, a peculiar fragrance that I love. Sea smell is the smell of home. I was back in Corpus Christi, a guest in the city of my childhood."

Commercial odors serve as locator beacons on the smellscape. For fifty-five years the Life Saver factory poured fruity sweetness over Port Chester, New York. The Mars candy plant keeps Hackettstown, New Jersey, smelling chocolaty, and the Maxwell House roasting operation periodically gives Hoboken a jolt of joe. A Snapple bottling plant fruitifies part of Baltimore, while a rendering plant, vinegar distillery, and giant bakery define other areas of town. McCormick & Co. blew a potpourri of spice across Baltimore for more than a century before relocating to Hunt Valley. A paper mill leaves a big, if unfavorable, impression on Muskegon, Michigan, and the Owens Country Sausage plant gives Sulphur Springs, Texas, a special yumminess.

WE COULD FILL an almanac with the site-specific scents of America. Because I grew up there, my nasal circuits are hopelessly imprinted on California. It's the source of dozens of characteristic smells—all true and equally essential—enough to fill a wing of the smell museum. The Golden State overwhelmed the intrepid Helen Keller: "I think I could write a book about the rich, warm, varied aromas of California; but I shall not start on that subject. It would take too long."

I'll give it a try. Start with the redwoods and the Sierra foothills full of *kit-kit-dizze* and coyote mint. Leave space for the La Brea Tar Pits and the pleasant, clean, tarry note that hovers over them. Include the stinkpots of Mount Lassen in the far north, and the sulfury hot springs of Esalen, down near Big Sur. The Pacific Coast has its own special collection: heaps of rotting kelp and the rich funk of tidal mud inside the Golden Gate. Depending on the wind direction, there's the stink of guano off Seal Rocks or the stench of the elephant seals at Point Año Neuvo.

The journalist and social observer Heather MacDonald grew up in the tony Bel Air section of Los Angeles. Living in a dense urban metropolis, she delighted in the nearby outdoors—a typical California contrast. "I spent a lot of time in the Santa Monica Mountains. The smell of the dry chaparral in the summer time and the eucalyptus and the wild mustard plants and the light. . . . There are so many smells that I associate with the land around here."

Eucalyptus, that Australian import, is everywhere in California. Another Australian, the Victorian box tree, has become part of the Southern California smellscape. Its nighttime perfume—an intoxicating blend of orange and honey—blankets Los Angeles every February. The local columnist Mary McNamara writes, "Seeping in through open windows, under doors, the scent saturates the air, the bedclothes, so dense you can taste it. Ambrosia rising, within and without."

The best way to sample California smells is by car. Drive down I-80 with the windows open as you pass the oil refineries in Pinole. Cruise past the Harris Ranch and the stockyards off I-5 in Coalinga and get the full blast of the cattle. Take US 101 through Gilroy and inhale the garlic. (And don't forget that the famous Lockheed "Skunk Works" in Los Angeles was named for the obnoxious smell of a nearby plastics factory.)

Maybe Helen Keller was right—California demands a lot of the cataloger, and these are just the bigger features of the smellscape. Zoom in to the level of neighborhoods and the picture gets more detailed, and even more evocative. Odor mapping is an exhausting effort. Is it really necessary to capture and preserve all this stuff that's just out there, floating around? Of course it is. The Hunt's tomato cannery in Davis is shuttered; the garlic depot in Vacaville is gone; Cannery Row smells only on paper; and it's a rare day when Fisherman's Wharf smells of a fresh catch. The recent past—our very lifetimes—is evaporating day by day.

CHAPTER 12

Our Olfactory Destiny

They were, I now saw, the most unearthly creatures it is possible to conceive. They were huge round bodies—or, rather, heads—about four feet in diameter, each body having in front of it a face. This face had no nostrils— indeed, the Martians do not seem to have had any sense of smell.

—H. G. WELLS, *The War of the Worlds*

IN THE IMAGINATION OF H. G. WELLS, MARTIANS WERE more advanced than humans: they didn't need a primitive sensory system with nostril holes and wet mucous membranes. Martians were big-eyed, big-brained, and gutless, with squidlike tentacles instead of arms and legs. What these creatures lacked in biology they made up for with technology: they roamed the Earth in mechanical exoskeletons. Since *The War of the Worlds* appeared in 1898, science-fiction writers and alien abductees have insisted that space visitors are noseless. I remember an *Outer Limits* episode in which the hero, a radio station engineer, makes contact with a creature from the fourth dimension. The curious alien asks him about the function of those strange holes below his eyes.

Like the Freudians, futurists are quick to dismiss the sense of smell as an evolutionary dead end. They speculate that our noses will shrink and our smelling ability will devolve along with it. But is this really our fate? To peer into our olfactory future, we must look toward smelling machines and olfactory genes.

Unlike space aliens, electronic noses are already among us—the first commercial units were delivered around 1992, intended for use in quality control in the flavor and fragrance industries. An e-nose uses an array of chemical sensors to detect odor molecules, and pattern analysis software to distinguish between them. Early models were large boxes that sat in the laboratory; more-recent handheld versions resemble something the meter reader might carry. What sets the e-nose apart from other chemical detectors—like those that measure breath alcohol or warn of carbon monoxide—is that it responds to a broad range of molecules. (Smoke alarms that work on optical principles are even less specific, which is why they sometimes mistake steam or fine dust for smoke.) The chemical sensors of an e-nose can be made from all sorts of materials, with conducting polymers being a popular choice. A conducting polymer changes its electrical resistance in the presence of volatile molecules. Some versions respond to odor at concentrations near the limits of human perception. These polymers are sensitive but not sophisticated; they are basically chemical sponges with different absorbent qualities.

The usefulness of an e-nose depends on its software as much as its sensors. The software extracts a pattern from the sensor input using formidable statistical methods. Multiple sensors give the e-nose a big advantage over single-molecule detectors. In particular, they avoid the pitfall of cross-interference. Imagine a fart detector that works by responding to a single molecule, namely hydrogen sulfide. Embarrassingly, it would go off every time your mom makes some egg salad. In contrast, a broadband e-nose reads the hydrogen sulfide along with other molecules, and would be less likely to mistakenly insult the lady of the house.

How well does an e-nose actually perform? Does it have the

potential to take jobs away from humans? Early models were intensely hyped by their manufacturers, and when the devices failed to live up to expectations, customers were left with a lingering negative impression of the technology. The hype hasn't entirely disappeared. An informal test in 2006 concluded that one brand of consumer e-nose—a handheld, battery-operated model that detected spoiled meat using the amines released by contaminating bacteria—oversold both the accuracy and the benefits of the device.

In general, the practical skills of the e-nose are real but modest; they include telling whether two smells are the same or different. This simple talent is useful in quality control where a manufacturer needs to keep batch-to-batch variation within limits or reject tainted raw materials. An e-nose excels at same/different judgments, and unlike human sensory panelists, it doesn't get tired or bored. (This doesn't mean it's maintenance-free; e-noses have to be recalibrated frequently owing to "sensor drift.") E-noses are good for dirty and dangerous jobs that humans don't want, such as monitoring emissions from animal feed lots and sewage treatment plants, or searching for land mines.

The e-nose also has a future in medicine. One device can detect diabetes from volatiles in the breath of a patient; another can find evidence of lung cancer. (Those cancer-sniffing dogs might be out of work before they know it.) An e-nose diagnostic scan would be quick and noninvasive. The main technical challenge is detecting a disease-related odor signal against a varying background of body odor.

Potential consumer applications could be in the offing, such as monitoring ambient fragrance levels—built-in scent systems for homes and offices will be more attractive if they include a feedback mechanism. A programmable olfactostat would maintain a pleasant level of scent in your environment; a wearable one could gauge the odor levels on your person.

Executives in the fragrance and flavor business dream of an e-nose that could stand in for a consumer test panelist. The device would be programmed with the exact preferences of urban preteens or suburban soccer moms in different zip codes. When presented with a test sample, it would respond "I like it" or "it's too floral." A robo-consumer has many advantages over human panelists: it's always on time and you don't have to pay it.

A surprising number of scientists are working on smell-capable robots; one of them published an entire book on the topic in 1999. Amy Loutfi, a researcher at Sweden's University of Örebro, has attached an e-nose to an intelligent, mobile robotic system. Her prototype resembles a Roomba—it wanders around an apartment under its own control, locating and identifying smells in the air. Loutfi improved her nose-bot's performance by adding psychological context to its decision-making process. The device identifies smells better when it knows it's in the living room rather than the bathroom.

Will police departments deputize the e-nose for remote drug sniffing? The U.S. Supreme Court held that thermal imaging of a suspected marijuana grower's home, because it relies on sense-enhancing technology that is not "in general public use," is an unconstitutional invasion of privacy. Under this standard, waving an e-nose downwind of a suspected grow house would also violate the Fourth Amendment's guarantee against unreasonable search and seizure. Until e-noses are available at Circuit City, police officers are going to have to rely on their own noses.

As with all technology, the law of unanticipated consequences will undoubtedly affect how the commercial e-nose market develops. For example, one near-term application is a pocket-size sniffer that tells from a woman's breath whether she is ovulating. The Ovulatron 5000 certainly will be a boon to couples trying to conceive, but it might also become a must-have technology for single guys on the prowl.

YOU CAN'T EXPECT an e-nose to work as soon as you take it out of the box. Training is essential, even to achieve competence at a simple same/different task. If its job is to pick out rotten apples, you must fill its database with examples of good apples and bad apples, so that it can create a statistical profile for each, and a decision-rule for telling them apart. An untrained e-nose would probably group wine samples according to alcohol content. It must be trained to distinguish Pinot Noir from Zinfandel. An e-nose is only as impressive as the training it gets. You can't follow your e-nose—you have to lead it.

An electronic sensing device appeals to hard-boiled process engineers because it is "objective." It frees them from discussions with sensory experts, and from dealing with emotional consumer panelists, at least in theory. But wait until the e-nose in Manufacturing gives a different reading than the one in Quality Control. Who does the engineer believe then? Good luck finding an objective way to settle that argument.

One thing our brain does very well is separate signal from noise. We can, for example, follow a single conversation at a cocktail party full of chattering voices. Similarly a perfumer can work in an office reeking of background smells that change from day to day. But tracking a target against ever-changing background odors is hard for an e-nose. Even harder is following a moving target against such a background: a ripening peach in a farmer's market, for example. Until it solves the cocktail-party problem, the e-nose will not be serious competition for the human nose.

AS TECHNOLOGY ADVANCES, the line between biology and hardware starts to blur. A group in Britain has developed what it calls "a truly biomimetic olfactory microsystem" by creating an artificial olfactory mucosa. In other words, they embedded electronic sensors in synthetic snot—a 10-micron-thick layer of an odor-retentive polymer called Parylene C. By delaying the detection of incoming odor molecules, the polymer slows the response time of the artificial nose, making it perform more like a biological one.

At the leading edge of technology, biological tissue is used as the odor sensor. For example, researchers can insert a mammalian odor receptor gene into yeast cells, which then manufacture the receptor and install it on their own cell surface. A tiny shred of the yeast cell membrane—including an intact, functioning receptor—is cut out and anchored to a chip that produces an electronic signal whenever the receptor is activated.

In a different approach, researchers use bacteria cells to produce odor receptors and then paint receptor-laden cell membrane fragments onto a tiny quartz crystal. The vibrational frequency of the crystal changes along with the weight of the layer coating it; this setup—known as a quartz crystal microbalance—is so sensitive it can tell when the receptors in the layer of bioslime have latched on to odor molecules, increasing its weight. An English company is using this technology to detect explosives. Another group has gone further and integrated entire rat olfactory cells into a semiconductor chip. They call this setup an olfactory neurochip, but it's really a rat-machine hybrid.

University-based scientists in France have pushed hybridism a step further: they have inserted a human odor receptor gene into yeast cells, which then express functioning human receptors for the odor molecule helional. The modified yeast cells become biosensors for helional. This is a technologically elegant but somewhat disturbing achievement: a combination of human DNA controlled by a foreign organism, which in turn is enslaved to a machine. Is this really a direction we want to pursue?

At some point in the development of these fusions of silicon and biology, the question becomes not whether the e-nose can replace the human nose, but whether we want it to. Would I let an e-nose sniff-scan me for lung cancer? Sure. Would I use a robotic odor sentinel? Maybe, especially if I had a BO problem. But do I really want my refrigerator to tell me, "I'm sorry, Avery, I can't let you eat those cold cuts"?

Genes of Scent

Supermarket tomatoes have no flavor. It's a common complaint, and a valid one. Commercial tomato varieties have less sugar, acid, and aroma than the wild type. On the other hand, they have better color, yield, disease resistance, and physical toughness, or what growers like to call "shippability." (Tomatoes are picked while still hard and green, to help them survive the trip to the store.) The guiding principle for tomato breeders is that it is better to look good than to taste good.

Help may be on the way: as scientists decipher the genetics of flavor chemical production in plants, they open the door to bioengineered flavor enhancement. One research group has discovered genes for the enzymes that are the first step in the biochemical production of phenylethyl alcohol, a key ingredient in tomato aroma. When overexpressed in transgenic tomato plants, these genes give the fruit ten times more rose alcohol, making it more fragrant than the ordinary variety. Another scientific team recently created a tastier tomato by altering the gene controlling a key enzyme involved in aroma production. They took the enzyme gene from lemon basil and inserted it into a tomato plant, where it modified biochemical activity to produce higher levels of key aroma molecules. This is cool science, but the proof of the pudding is in the eating, and here the new transgenic tomato is a winner: It was preferred by panelists in taste tests.

Rather than import genes from other plant species, genetic

engineers may decide to pluck useful ones from so-called heirloom tomatoes, the distinctive-looking and interesting-tasting varieties prized in farmers markets across the country. Heirloom tomatoes—with names like Marvel Striped and Purple Cherokee—existed before the breeding programs that created the standardized, disease-resistant, high-yield, shippable kind that dominate today's supermarket shelves. Ann Noble, the UC Davis wine expert and creator of the Wine Aroma Wheel, has been lured out of retirement by Central Valley tomato growers looking to promote their heirloom business. They hope she will do for tomatoes what she did for wine—encourage sensory analysis to help consumers understand and appreciate all their varied aromatic qualities.

If there is anything more dispiriting than a flavorless tomato, it is a scentless rose. Along with chrysanthemums, tulips, lilies, and carnations, roses are the top sellers in the cut-flower market, with worldwide sales estimated at $40 billion a year. Where has all the fragrance gone? There are more than a hundred species of roses, yet most of those in commercial production result from crosses between only eight species. Like tomatoes, these varieties were not selected for fragrance, but for traits that the cut-flower industry prizes: flower color and shape, yield, vase life, and resistance to insects and disease.

Perfume chemists analyze floral scents down to the last molecule, but it's not their job to find out how plants make the scent in the first place. Nor were academic researchers interested: in 1994, not a single floral scent enzyme had been identified. Then the biologist Eran Pichersky began to study a native California wildflower known as Brewer's Clarkia. This unusual species—a night-blooming, moth-pollinated evening primrose—grows in only the San Francisco Bay Area. Pichersky's team chemically characterized its scent and found that one ingredient—linalool—is produced by an enzyme called linalool synthase. When they successfully identified the gene that produced the enzyme, they opened up a whole new scientific field: floral scent biochemistry.

Since then, Pichersky and others have looked for the scent-producing enzymes in the Fragrant Cloud rose, and the genes that code for them. They hope to transfer those genes into a scentless rose like the Golden Gate cultivar.

Biotechnologists may ride to the rescue of rose scent. They have a toolbox full of techniques to transfer genes into plants. They can literally shoot new genes into plant cells using microscopic DNA-coated particles of gold or tungsten. Or they can use the microorganism *Agrobacterium* to install the genes for them. Not only can genetic engineers restore a plant's original scent, they can give it the scent of another species. It's a dizzying thought: roses that smell like violets, asters that smell like lilacs. The creation of transgenically fragrant flowers will be a victory for biotechnology and may ease public acceptance of biotech crops.

This would all seem like a perfect opportunity for the cut-flower industry. Yet Eran Pichersky tells me that producers are reluctant to make the effort. According to their market research, consumers claim that scent matters, but sales figures don't reflect it. Consumer choice is driven by color and visual appeal. In any case, most flowers are bought as gifts, which means the purchaser doesn't live with the scent, or lack thereof. Perhaps it is true, as Shakespeare said, that "to throw a perfume on the violet . . . Is wasteful and ridiculous excess."

The Genes of Perception

Imagine a DNA test in which a marketer predicts your fragrance preferences in ten minutes using a drop of your saliva. Rapid, saliva-based clinical diagnostics like home pregnancy tests are already in use. Why shouldn't there be point-of-sale diagnostics? Wouldn't you trade a little spit to find your perfect fragrance?

The person-to-person variability in odor perception is enormous. To get an idea of the scale, compare it to color vision. Imagine that instead of three kinds of color blindness there were dozens, and that each type affected up to 75 percent of the population instead of only 6 percent. Smell scientists struggle to explain this variability; it remains one of the biggest mysteries about the sense of smell. Why are some people able to smell a particular molecule and others not? Why do some people find it pleasant and others do not?

Cultural factors—the favorite explanation of academic researchers —certainly play a role in odor preferences. But cultural explanations don't go too far in explaining the extensive differences between people within the same culture. Biological factors, which receive surprisingly little attention, may account for much of this variation. For example, certain specific anosmias—the inability of a person with otherwise normal smell to detect a specific type of molecule— have a biological basis, namely the lack of a receptor for the molecule in question. There are a couple of dozen specific anosmias, but they account for merely a fraction of the total variation in odor perception.

The key to the mystery may reside more broadly in the human genome. A tantalizing possibility is that your olfactory receptor genes determine how you smell the world, and why you smell it differently than other people. Everyone has roughly 350 olfactory receptors, but not necessarily the same 350 as the next person. In addition, the gene for a given receptor can show subtle variation in DNA sequence from person to person.

The science of genetics links genotype (a person's DNA profile) to phenotype (a person's physical and mental traits). Several laboratories around the world are exploring the genetics of odor perception. Their first challenge is to characterize a person's odor perception phenotype—in other words, to measure the sensitivity to, and preference for, a wide range of smells. The next step is to use DNA analysis to establish a person's odor receptor genotype. Researchers expect that people with similar phenotypes have certain genetic traits in common. For example, people who like musk, hate

grape, and are indifferent to patchouli may have certain odor receptor variants in common, and these biomarkers could become the
basis of the in-store perfume preference diagnostic.

The first step toward a functional genomics of olfaction has
already been taken. Researchers at Rockefeller and Duke Universities have discovered that variations in one odor receptor gene are
responsible for differences in how people perceive the molecules
called androstenone and androstadienone. These genetic variations,
known as single nucleotide polymorphisms, have the effect of mut-
ing the intensity and unpleasantness of these two smelly molecules.
It's astounding that such tiny mutations can have such major consequences for odor perception. Yet this is just the tip of the iceberg—we
can expect many more examples in the years ahead.

Knowing the link between genes and odor perception will profoundly change how we think about smell. Pavlovian learning and
Proustian remembering will have to share the stage with biology. The
discovery of biological markers for scent preference would revolution-
ize the design and marketing of fragrance. Instead of making products
that appeal to the market as a whole (and satisfy no one in particular),
perfumers could target scents to biologically defined market segments.
A perfumer designing something for the musk-loving, grape-hating,
patchouli-indifferent audience will have a tremendous advantage over
a competitor working with the old hit-or-miss method.

THE GENOMIC AGE of odor perception will be exciting. We will be
able to alter odor perception at a fundamental biological level—
enhancing the response of a receptor, for example, or blocking it
from working at all. These molecular-level interventions could lead
to new types of consumer products. Imagine a long-lasting nasal
spray for the medical staff in hospitals and nursing homes. One
squirt at the start of a shift would knock out the ability to smell the
ammonialike notes in urine, but leave the perception of other odors

unchanged. The product would work by stopping a specific class of molecules from triggering a sensation. A narrow-range odor blocker like this would make the hospital a more pleasant place to work; and happier staff make for happier patients. Think of all the other occupations—stockyard worker, plumber, refinery employee—that could benefit from selective molecular nose-filters.

Next, imagine a new kind of diet product—one with an immediate and profound effect on appetite: food would lose its appeal and odor-induced cravings would disappear. In biological terms this would be a wide-range odor blocker that interferes with many types of receptors. By reducing odor perception across the board, including food aroma, the blocker would help dieters stay on their program. A recent patent application makes such a claim for a calcium channel blocker— a type of drug usually used to control high blood pressure. Applied directly into the nose, it would temporarily stop the sensory cells from functioning, and reduce or abolish the user's ability to smell.

By changing receptor function in other ways, we may be able to enhance odor perception. Imagine a product that selectively boosts the perception of certain body odors, like your husband's pheromones. It might heighten sexual interest or arousal and be a useful treatment for sexual dysfunction. (It would probably become popular with ravers, clubbers, and swingers too—a nasal Ecstasy.) Another possibility is a broad-range odor booster. The results could be mind-blowing. The neurologist and essayist Oliver Sacks once described a patient who experienced heightened smell awareness while pumped up on amphetamines, cocaine, and PCP. The immediacy and clarity of smells was so great that he could find his way around New York by nose alone. Not everybody would want to have such a peak experience, although it's a product that Emily Dickinson would have paid top dollar for. At a lower dose, a broad-range odor booster might relieve smell impairment in the elderly. Their food will taste better, they will eat more, and their nutrition will improve. Who knows, it might even alleviate the psychological depression that creeps along in tandem with the sensory deprivation of old age.

The temporary tweaking of existing odor receptors is, from a biotechnologist's point of view, pretty straightforward. The sensory cells of the nose are in direct contact with the outside world, separated by only a thin layer of mucus. They can be reached easily with a topical nasal spray, which means a minimal amount of active ingredient and less chance of side effects. The really weird possibilities go deeper: imagine acquiring a new odor receptor gene. All you would have to do is take a big snort from spray bottle of genetically modified adenovirus, and within days you'd be having a new smell experience. Perhaps your specific anosmia to androstenone will be cured, enabling you for the first time to enjoy the expensive pleasure of truffles. Perhaps you will have a new, deeper appreciation of musky perfumes. Suppose the inhaled virus particles contained all the odor receptors a dog has and you haven't. By the weekend you'd be smelling things our species hasn't picked up in millions of years. The experience might be disconcerting at first, like getting powerful new contact lenses. Your brain would need time to adjust to the new odor input and bring it into focus.

This is a fantasy, but not a completely implausible one. Gene-transfer technology is routinely used in research labs. DNA is carried from one organism to another in a modified adenovirus—the virus that causes the common cold. The virus is unable to replicate on its own, but it can worm its way into the DNA of the host cells and trick them into reproducing the transferred gene.

Gene-transfer technology for humans is usually thought of in terms of treatment for life-threatening illness. But in the spirit of William Gibson's *Neuromancer*, where characters favor trans-species body modification, I predict it will be used first for nonmedical and entirely unnecessary aesthetic enhancements to the human body. In similar fashion, the first animal-to-human odor receptor implant will take place for kicks, not for cure.

Transspecies genetic engineering of sensory systems is already happening in the lab. Mice have been given new photoreceptor genes, and the sex pheromone receptor of the silkworm moth has been transferred into a fruit fly. One day we will be able to control our own olfactory destiny. What do you want to smell like?

—

> *The horizon's edge, the flying sea-crow, the fragrance of salt*
>   *marsh and shore mud,*
> *These became part of that child who went forth every day,*
>   *and who now goes, and will always go forth every day.*
>
>   —WALT WHITMAN, *Leaves of Grass*

Acknowledgments

For encouraging me to write this book in the first place, I thank Barbara Ivins and Mandy Aftel. For advice on how to go about it, and moral support during the writing of it, I thank Tom Higgins and Lisa Verge Higgins, and my excellent agent, Michelle Tessler. The expert guidance of my editor, Lucinda Bartley, made this a better book.

I benefited from the insights and recollections of the many people I interviewed. For helping me reconstruct the history of Smell-O-Vision and AromaRama, I thank Carmen Laube, Novia Laube, Glenda Jensen, Hal Williamson, Ronnie Reade, Luz Gunsberg, Paul Baise, John Waters, Mark Gulbrandsen, James Bond, Steve Kraus, and Denise Garrity. For sharing with me their scientific and technical expertise on olfactory matters, I thank Kari Arienti, René Morgenthaler, Terry Acree, Eric Berghammer, Eran Pichersky, Steven Sunshine, James Woodford, and Paul Breslin. Over the years I have benefited from discussions of smelly topics with Paul Rozin, Gary Beauchamp, and Michael O'Mahoney. Roman Kaiser deserves special thanks for helping me track down the molecule that inspired John Muir in the Sierra Nevada so long ago.

I am grateful to everyone who gave me a hand here or there; without these graciously provided assists, I could not have completed this work. These friends and colleagues include Gregg Rapaport, Owen Brown, Felix Aeppli, Jeff Freda, Dennis Passe, Jim

Walker, Jennifer Stevenson (and her pals Marge Kriz, Alan Rafaelson, and Herb Kraus), Peter F. Stucki, Bernadette Meier, Jeanine Delwiche, Charlotte Tancin, Ernest Sanders, Candace Jackson, Laurence Dryer, John Lundin, Kat Anderson, Gordon Shepherd, Mark B. Adams, Gunnar Broberg, John Canemaker, Betty Gilbert, Tirza True Latimer, John Prescott, Harris Jones, Mark Greenberg, Steve Jellinek, Marci Pelchat, Alan Fridlund, John Steele, Sue Van Inwegen, Leti Bocanegra, Tom Rigney, Carol Christensen, Steven Mintz, Melissa Mintz, Diana Hollander, Lilly Hollander, Larry Clark, Jim Bolton, Stephen Porter, Rolf Bell, and Alison Cocotis.

I am grateful to the many colleagues who generously provided copies of their publications and data: Pam Dalton, Jim Drobnick, Bob Frank, Hildegarde Heymann, Devon Hinton, David Hornung, Thomas Hummel, David Laing, Zachary F. Mainen, Florian Mayer, Yoshihito Niimura, Ann Noble, Tim Pearce, Timothy D. Smith, Eric Spangenberg, Dick Stevenson, Denise Tieman, Omer Van den Bergh, Bill Wood, Don Wright, Christina Zelano, Lorie Fulton, Debra Zellner, Joël Candau, John D'Auria, Claire Murphy, Bryan Raudenbush, Ralph Both, and Kirsten Sucker.

Finally, and most importantly, I thank my wife, Susanne, and our beautiful daughters, Alice and Lydia, for their patience and love and support.

Notes

Chapter 1. Odors in the Mind

1 *"It is very obvious"* Alexander Graham Bell, "Discovery and invention," *National Geographic*, June 1914.

1 *Some people aren't daunted* Michael Murphy, *The Future of the Body: Explorations into the Further Evolution of Human Nature* (Los Angeles: Jeremy P. Tarcher, Inc., 1992), p. 68; Vitus B. Dröscher, *The Magic of the Senses: New Discoveries in Animal Perception* (New York: E. P. Dutton, 1969), p. 100 (translation of 1966 German original); Andrew Hamilton, "What science is learning about smell," *Science Digest*, November 1966, p. 81–84.

2 *I began with a paper* M. Milinski and C. Wedekind, "Evidence for MHC-correlated perfume preferences," *Behavioral Ecology* 12 (2001): 140–49; B. C. Prasad and R. R. Reed, "Chemosensation: Molecular mechanisms in worms and mammals," *Trends in Genetics* 15 (1999): 150–53; Trygg Engen, *The Perception of Odor*, (Academic Press, 1982), p. 99; R. H. Wright, *The Science of Smell* (London: George Allen & Unwin, 1964), p. 80.

3 *I found it once more* John M. deMan, *Principles of Food Chemistry*, 3rd edition (New York: Springer, 1999), p. 287; A. Dravnieks, "Current status of odor theories," in *Flavor Chemistry*, edited by Irwin Hornstein (Washington, DC: American Chemical Society, 1966); R. M. Hainer, A. G. Emslie, and A. Jacobson, "An information theory of olfaction," *Annals of the New York Academy of Sciences* 58 (1954):158–74.

3 *he was Ernest C. Crocker* "Crocker speaks at initiation banquet," *The Tech* (MIT student paper), May 15, 1934, p. 1; "Ernest Charlton Crocker (1888–1964)," in *A Dictionary of Psychology*, edited by Andrew M. Colman (Oxford University Press, 2001).

3 *Crocker and another* E. C. Crocker and L. F. Henderson, "Analysis and

classification of odors: An effort to develop a workable method," *The American Perfumer & Essential Oil Review* 22 (1927):325; Edwin G. Boring, "A new system for the classification of odors," *American Journal of Psychology* 40 (1928):345–49.

4 *took the number and ran with it*   ADL continues to cite Crocker and his big numbers to this day. See "Sensory Benchmarking: The U.S. Soymilk Market 2001," a report by Soyatech, Inc., and Arthur D. Little, Inc., p. 5.

4 *A pair of industrial engineers*   W. Barfield and E. Danas, "Comments on the use of olfactory displays for virtual environments," *Presence* 5 (1996): 109–21.

4 *naming categories of color*   T. Regier, P. Kay, and N. Khetarpal, "Color naming reflects optimal partitions of color space," *Proceedings of the National Academy of Sciences USA* 104 (2007):1436–41.

6 *Noble's approach to aroma*   H. Heymann and A. C. Noble, "Descriptive analysis of commercial Cabernet Sauvignon wines from California," *American Journal of Enology and Viticulture* 38 (1987):41–44.

7 *Wine Aroma Wheel*   A. C. Noble, R. A. Arnold, et al., "Modification of a standardized system of wine aroma terminology," *American Journal of Enology and Viticulture* 38 (1987):143–46; "foodstuffs" quote from p. 144. The Wine Aroma Wheel can also be found online at www.winearomawheel. com.

8 *smell classification for beer*   M. C. Meilgaard, "Flavor Chemistry of Beer," *Master Brewers Association of the Americas Technical Quarterly* 12 (1975):107.

9 *appeal of aroma wheels*   N. P. Jolly and S. Hattingh, "A brandy aroma wheel for South African brandy," *South African Journal for Enology and Viticulture* 22 (2001):16–21; G. A. Burlingame, I. H. Suffet, et al., "Development of an odor wheel classification scheme for wastewater," *Water Science and Technology* 49 (2004):201–9. Mandy Aftel's Natural Perfume Wheel can be found at www.aftelier.com.

10 *there are at least 1,000*   Nancy Jeffries, "Fragrance Awards, Beauty Trends and the Scent of Peace," *GCI*, June 2006, pp. 20–22; Jeff Falk, "How's that for originality?" *GCI*, October 2005, p. 4.

10 *professional perfumers*   Robert R. Calkin and J. Stephan Jellinek, *Perfumery: Practice and Principles* (New York: John Wiley & Sons, 1994), p. 24.

11 *The leading teaching technique*   Interviews with Kari Arienti, June 30, 2004 and December 20, 2007, and with René Morgenthaler, December 19, 2007.

12 *to think like a perfumer*   Calkin and Jellinek, *Perfumery*, pp. 24, 61.

15 *"Vibrantly feminine floralcy"*   The Fragrance Foundation Reference Guide, 23rd edition (2002), p. 64.

15 *"It's intended to target"*   Global Cosmetic Industry, April 2004, p.14.

16 "Après l'Ondée *evolves only slightly"*   Luca Turin, quoted in Chandler Burr, *The Emperor of Scent* (New York: Random House, 2002), p. 36.

17 *"This is the scent of the darkness"*   Chandler Burr, "Dark Victory," *New York Times "T" Style Magazine*, August 27, 2006.

17 *perfume fans might prefer*   "Now that's stinking rich," *Sunday Metro*, July 13, 2006 (published online as well).

18 *bookish desk-jockey*   Ernst Mayr, *The Growth of Biological Thought: Diversity, Evolution, and Inheritance* (Boston: Harvard University Press, 1982).

18 *the actual treatise*   Carl von Linné, *Odores medicamentorum*, Amoenitates Academicae 3, 183–201, 1752. The specific author is Andreas Magnus Wåhlin, series no. XXXVIII. Little is known about Wåhlin, but scholars believe Linnaeus was the driving force behind the smell project. (From correspondence with Gunnar Broberg.)

18 *To his way of thinking*   To get the flavor of Linnaeus's odd logic, see Aphorhisms 358–362 in his *Philosophia Botanica*, translated by Stephen Freer (Oxford University Press, 2003).

19 *wild-goose chase*   Modern researchers continue to mistake Linnaeus for an odor classifier. See, for example, M. Zarzo and D. T. Stanton, "Identification of latent variables in a semantic odor profile database using principal component analysis," *Chemical Senses* 31 (2006):713–724.

19 *Zwaardemaker's classification*   Hendrik Zwaardemaker, *Die Physiologie des Geruchs* (Leipzig: W. Engelmann, 1895).

19 *German physiologist Hans Henning*   Hans Henning, *Der Geruch* (Leipzig: J. A. Barth, 1916).

20 *dismantling of the odor prism*   For background on this, see M. W. Levine, *Fundamentals of Sensation and Perception*, 3rd edition (Oxford, New York: Oxford University Press, 2001); P. M. Wise, M. J. Olsson, and W. S. Cain, "Quantification of odor quality," *Chemical Senses* 25 (2000):429–43; Stanley Finger, *Origins of Neuroscience: A History of Explorations into Brain Function* (Oxford, New York: Oxford University Press, 2001).

20 *new system of smell classification*   Crocker and Henderson, "Analysis and classification of odors," p. 325.

21 *a stunning blow*   S. Ross and A. E. Harriman, "A preliminary study of the Crocker-Henderson odor-classification system," *American Journal of Psychology* 62 (1949):399–404.

21 *burst of innovation*   John E. Amoore, *Molecular Basis of Odor* (Springfield, Ill.: Thomas, 1970).

23 *Australian psychologist David Laing*   Key findings of Laing's that I discuss here are referenced in A. Livermore and D. G. Laing, "The influence of chemical complexity on the perception of multicomponent odor mixtures," *Perception & Psychophysics* 60 (1998):650–61, and in A. Livermore and D. G. Laing, "The influence of odor type on the discrimination and identification of odorants in multicomponent odor mixtures," *Physiology & Behavior* 65 (1998):311–20.

Chapter 2. The Molecules That Matter

25 *The planet Mars*   Leonard David, "Mars stinks: Sulfur deposits may make Red Planet putrid," Space.com/scienceastronomy/mars_stinks_040308.html (March 8, 2004).

28 *researchers in Salt Lake City*   J. G. Moore, L. D. Jessop, and D. N. Osborne, "Gas-chromatographic and mass-spectrometric analysis of the odor of human feces," *Gastroenterology* 93 (1987):1321–29.

28 *olfactory analysis of farts*   F. L. Suarez, J. Springfield, and M. D. Levitt, "Identification of gases responsible for the odour of human flatus and evaluation of a device purported to reduce this odour," *Gut* 43 (1998): 100–104.

29 *chemical composition of baby farts*   T. Jiang, F. L. Suarez, et al., "Gas production by feces of infants," *Journal of Pediatric Gastroenterology and Nutrition* 32 (2001):534–41.

29 *Another cherished belief*   T. I. Case, B. M. Repacholi, and R. J. Stevenson, "My baby doesn't smell as bad as yours: The plasticity of disgust," *Evolution and Human Behavior* 27 (2006):357–65.

30 *outsized cultural impact*   "Blagojevich Enjoys Campaign Trail, While Ryan Endures It: Democrat's Baggage Hasn't Slowed His Momentum," *St. Louis Post-Dispatch*, October 6, 2002; Warhol attribution from www .hempfiles.com.

31 *are therefore odorless*   The National Toxicology Program (U.S. Department of Health and Human Services) lists the odor of THC as "none found," an impression confirmed by Eran Pichersky (interview September 19, 2006)

and by Jim Woodford (interview April 11, 2007). THC, cannabinol, and cannabidiol are "nonvolatile" according to M. Rothschild, G. Bergström, and S. Wängberg, "*Cannabis sativa*: Volatile compounds from pollen and entire male and female plants of two variants, Northern Lights and Hawaiian Indica," *Botanical Journal of the Linnean Society* 147 (2005):387–97.

31 *Early in his career* Interview with W. James Woodford, April 11, 2007.

31 *found in flower scents* J. Horiuchi, D. V. Badri, et al., "The floral volatile, methyl benzoate, from snapdragon (*Antirrhinum majus*) triggers phytotoxic effects in *Arabidopsis thaliana*," *Planta* 226 (2007):1–10; M. Kondo, N. Oyama-Okubo, et al., "Floral scent diversity is differently expressed in emitted and endogenous components in *Petunia axillaris* lines," *Annals of Botany* (London) 98 (2006):1253–59.

32 *fake Ecstasy aroma* N. Lorenzo, T. Wan, et al., "Laboratory and field experiments used to identify *Canis lupus* var. *familiaris* active odor signature chemicals from drugs, explosives, and humans," *Analytical and Bioanalytical Chemistry* 376 (2003):1212–24.

32 *chemistry of marijuana* L. V. Hood, M. E. Dames, and G. T. Barry, "Headspace volatiles of marijuana," *Nature* 242 (1973):402–3; L. V. Hood and G. T. Barry, "Headspace volatiles of marihuana and hashish: Gas chromatographic analysis of samples of different geographic origin," *Journal of Chromatography* 166 (1978):499–506; S. A. Ross and M. A. ElSohly, "The volatile oil composition of fresh and air-dried buds of *Cannabis sativa*," *Journal of Natural Products* 59 (1996):49–51; Rothschild, Bergström, and Wängberg, "*Cannabis sativa*," pp. 387–97.

33 *Beck concert* Matt Coker, "Live Review: Smell You Later; Beck: The Pacific Amphitheatre," *OC Weekly*, vol. 10, no. 47, July 29–August 4, 2005.

34 *an anomalous finding* R. O. Pihl, D. Shea, and L. Costa, "Odor and marijuana intoxication," *Journal of Clinical Psychology* 34 (1978):775–79.

34 *industry pundits were coy* *Multichannel News*, August 14, 2006; "Pass the Scent Strip," Ad Age.com, August 10, 2006.

35 *"The air was steaming"* John Muir, *The Overland Monthly*, June 1875.

36 *the plant he smelled* Karen Wiese, *Sierra Nevada Wildflowers* (Helena, Montana: Falcon Publishing Inc., 2000).

36 *"approximately the amount given off"* R. Kaiser, "Scents from rain forests," *Chimia* 54 (2000):346–63.

37 *study and preserve the scent* R. Kaiser, "Vanishing flora—lost chemistry:

The scents of endangered plants around the world," *Chemistry & Biodiversity* 1 (2004):13–27.

38 *Hexenone has been fingered*   S. Widder, A. Sen, and W. Grosch, "Changes in the flavour of butter oil during storage; identification of potent odorants," *Zeitschrift für Lebensmitteluntersuchung und –Forschung A* 193 (1991):32–35; I. Blank, K.-H. Fischer, and W. Grosch, "Intensive neutral odourants of linden honey: Differences from honeys of other botanical origin," *Zeitschrift für Lebensmitteluntersuchung und –Forschung A* 189 (1989):426–33; D. D. Roberts and T. E. Acree, "Effects of heating and cream addition on fresh raspberry aroma using a retronasal simulator and gas chromatography olfactometry," *Journal of Agricultural and Food Chemistry* 44 (1996):3919–25.

40 *"strong and distinct" odor*   George B. Longstaff, *Butterfly-hunting in Many Lands* (London: Longmans, Green, and Co., 1912), p. 491.

41 *Males of the Green-veined White*   J. Andersson, A. K. Borg-Karlson, C. Wiklund, "Sexual conflict and anti-aphrodisiac titre in a polyandrous butterfly: Male ejaculate tailoring and absence of female control," *Proceedings: Biological Sciences* 271 (2004):1765–70; J. Andersson, A. K. Borg-Karlson, and C. Wiklund, "Antiaphrodisiacs in pierid butterflies: A theme with variation!" *Journal of Chemical Ecology* 29 (2003):1489–99.

41 *countermeasures can backfire*   N. E. Fatouros, M. E. Huigens, et al., "Chemical communication: Butterfly anti-aphrodisiac lures parasitic wasps," *Nature* 433 (2005):704.

41 *dead-horse arum fakes the stench*   M. C. Stensmyr, I. Urru, et al., "Pollination: Rotting smell of dead-horse arum florets," *Nature* 420 (2002):625–26.

42 *An Australian orchid*   F. P. Schiestl, R. Peakall, et al., "The chemistry of sexual deception in an orchid-wasp pollination system," *Science* 302 (2003):437–38.

43 *While prospecting there*   Kaiser, "Scents from rain forests," p. 350.

44 *hyperlink from molecule to substance*   P. K. Ong and T. E. Acree, "Gas chromatography/olfactory analysis of lychee (*Litchi chinesis* Sonn.)," *Journal of Agricultural and Food Chemistry* 46 (1998):2282–86; P. K. Ong and T. E. Acree, "Similarities in the aroma chemistry of Gewurztraminer variety wines and lychee (*Litchi chinesis* Sonn.) fruit," *Journal of Agricultural and Food Chemistry* 47 (1999):665–70; R. Triqui and N. Bouchriti, "Freshness assessments of Moroccan sardine (*Sardina pilchardus*): Comparison of overall sensory changes to instrumentally determined volatiles," *Journal of Agricultural and Food Chemistry* 51 (2003):7540–46; F. Piveteau, S. Le Guen, et al. "Aroma of fresh oysters *Crassostrea gigas*: Composition and aroma

note," *Journal of Agricultural and Food Chemistry* 48 (2000):4851–57; K. Fukami, S. Ishiyama, et al., "Identification of distinctive volatile compounds in fish sauce," *Journal of Agricultural and Food Chemistry,* 50 (2002):5412–16; L. R. Freeman, G. J. Silverman, et al., "Volatiles produced by microorganisms isolated from refrigerated chicken at spoilage," *Applied and Environmental Microbiology* 32 (1976):222–31.

45 *fewer than 5 percent* W. Grosch, "Evaluation of the key odorants of foods by dilution experiments, aroma models and omission," *Chemical Senses* 26 (2001):533–45.

46 *Using aroma models* M. Czerny, F. Mayer, and W. Grosch, "Sensory study on the character impact odorants of roasted arabica coffee," *Journal of Agricultural and Food Chemistry* 47 (1999):695–99.

46 *Livestock feeding operations* E. A. Bulliner, J. A. Koziel, et al., "Characterization of livestock odors using steel plates, solid-phase microextraction, and multidimensional gas chromatography-mass spectrometry-olfactometry," *Journal of the Air & Waste Management Association* 56 (2006):1391–1403; D. W. Wright, D. K. Eaton, et al., "Multidimensional gas chromatography-olfactometry for the identification and prioritization of malodors from confined animal feeding operations," *Journal of Agricultural and Food Chemistry* 53 (2005):8663–72.

47 *As beets are processed* P. Pihlsgard, M. Larsson, et al., "Volatile compounds in the production of liquid beet sugar," *Journal of Agricultural and Food Chemistry* 48 (2000):44–50.

Chapter 3. Freaks, Geeks, and Prodigies

49 *people are not accurate* C. M. Philpott, C. R. Wolstenholme, et al., "Comparison of subjective perception with objective measurement of olfaction," *Archives Otolaryngology—Head & Neck Surgery* 134 (2006):488–90.

49 *formally recognized* *Federal Register* 71, no. 109, p. 32834, June 7, 2006.

49 *1 to 2 percent of the U.S. population* B. A. Nguyen-Khoa, E. L. Goehring, et al., "Epidemiologic study of smell disturbance in 2 medical insurance claims populations," *Archives of Otolaryngology—Head & Neck Surgery* 133 (2007): 748–57.

50 *It takes very little force* J. R. de Kruijk, P. Leffers, et al., "Olfactory function after mild traumatic brain injury," *Brain Injury* 17 (2003):73–78.

50 *recovery can take months* J. Reden, A. Mueller, et al., "Recovery of olfactory function following closed head injury or infections of the upper respiratory

tract," *Archives of Otolaryngology—Head & Neck Surgery* 132 (2006):265–69; R. Harris, T. M. Davidson, et al., "Clinical evaluation and symptoms of chemosensory impairment: One thousand consecutive cases from the Nasal Dysfunction Clinic in San Diego," *American Journal of Rhinology* 20 (2006):101–8.

50   *psychologically devastating*   T. Hummel and S. Nordin, "Olfactory disorders and their consequences for quality of life," *Acta Oto-Laryngologica* 125 (2005):116–21; E. H. Blomqvist, A. Bramerson, et al., "Consequences of olfactory loss and adopted coping strategies," *Rhinology* 42 (2004):189–94.

51   *little data to suggest*   D. V. Santos, E. R. Reiter, et al., "Hazardous events associated with impaired olfactory function," *Archives of Otolaryngology—Head & Neck Surgery* 130 (2004):317–19.

51   *young English anosmic*   Lucy Mangan, "Scents and sensitivity," *The Guardian*, July 20, 2004.

51   *reporter who is smell-blind*   Karen Ravn, "Sniff . . . and Spend: Now that the retail industry has caught a whiff of smells' success, prepare your nose for the marketing onslaught," and "First Person: Hey, there's no sense missing what you can't smell," *Los Angeles Times*, August 20, 2007.

51   *olfactory hallucinations*   M. S. Greenberg, "Olfactory hallucinations," in M. J. Serby and K. L. Chobor, eds., *The Science of Olfaction* (New York: Springer-Verlag, 1992), pp. 467–99.

52   *the condition is called parosmia*   B. N. Landis, J. Frasnelli, and T. Hummel, "Euosmia: A rare form of parosmia," *Acta Otolaryngologica* 126 (2006):101–3; W. B. Shelley and E. D. Shelley, "The smell of burnt toast: A case report," *Cutis* 65 (2000):225–26; P. Bonfils, P. Avan, et al., "Distorted odorant perception: Analysis of a series of 56 patients with parosmia," *Archives of Otolaryngology—Head & Neck Surgery* 131 (2005):107–12.

52   *persistent hallucinations*   C. Lochner, D. J. Stein, "Olfactory reference syndrome: Diagnostic criteria and differential diagnosis," *Journal of Postgraduate Medicine* 49 (2003):328–31.

53   *A German psychologist*   N. Klutky, "Geschlechtsunterschiede in der Gedächtnisleistung für Gerüche, Tonfolgen und Farben [Sex differences in memory performance for odors, tone sequences and colors]," *Zeitschrift für experimentelle und angewandte Psychologie* 37 (1990):437–46.

53   *"At least five time per week"*   Dave Barry, "The Nose Knows," *Miami Herald*, May 5, 1998.

53   *brain structures*   A. Garcia-Falgueras, C. Junque, et al., "Sex differences in the human olfactory system," *Brain Research* 1116 (2006):103–11.

53   *some male-female differences*   J. K. Olofsson and S. Nordin, "Gender differences in chemosensory perception and event-related potentials," *Chemical Senses* 29 (2004):629–37.

53   *higher verbal fluency*   M. Larsson, M. Lovden, and L. G. Nilsson, "Sex differences in recollective experience for olfactory and verbal information," *Acta Psychologica* 112 (2003):89–103.

54   *dramatic olfactory sex differences*   P. Dalton, N. Doolittle, and P. A. Breslin, "Gender-specific induction of enhanced sensitivity to odors," *Nature Neuroscience* 5 (2002):199–200.

54   *most remarkably*   Interview with Paul Breslin, July 31, 2007.

54   *within days of birth*   H. J. Schmidt and G. K. Beauchamp, "Human olfaction in infancy and early childhood," in Serby and Chobor, 1992, pp. 378–95.

54   *Tiger's view*   Lionel Tiger, *The Pursuit of Pleasure* (New York: Little, Brown, 1992), p. 64.

54   *rate of decline varies*   C. J. Wysocki and A. N. Gilbert, National Geographic Smell Survey, "Effects of age are heterogenous," *Annals of the New York Academy of Sciences* 561 (1989):12–28.

55   *a simple yes/no format*   J. Corwin, "Assessing olfaction: Cognitive and measurement issues," in Serby and Chobor, 1992, pp. 335–54.

55   *including several recent ones*   A. Knaapila, K. Keskitalo, et al., "Genetic component of identification, intensity and pleasantness of odours: A Finnish family study," *European Journal of Human Genetics* 15 (2007):596–602.

55   *"smoking did not reduce"*   A. Mackay-Sim, A. N. Johnston, et al., "Olfactory ability in the healthy population: reassessing presbyosmia," *Chemical Senses* 31 (2006):763–71.

55   *Smell Survey reported mixed results*   A. N. Gilbert and C. J. Wysocki, "The Smell Survey Results," *National Geographic* 172 (1987):514–25.

56   *"The lack of a statistically significant"*   A. Bramerson, L. Johansson, et al., "Prevalence of olfactory dysfunction: The Skovde population-based study," *Laryngoscope* 114 (2004):733–37.

57   *"I have not"*   Helen Keller, *The World I Live In* (New York: The Century Co., 1908).

57   *six studies have compared*   C. Murphy and W. S. Cain, "Odor identification: The blind are better," *Physiology & Behavior* 37 (1986):177–80; R. S. Smith, R. L. Doty, et al., "Smell and taste function in the visually impaired," *Perception & Psychophysics* 54 (1993):649–55; R. Rosenbluth, E. S. Grossman, and M. Kaitz, "Performance of early-blind and sighted children on olfactory

tasks," *Perception* 29 (2000):101–10; C. E. Wakefield, J. Homewood, and A. J. Taylor, "Cognitive compensations for blindness in children: An investigation using odour naming," *Perception* 33 (2004):429–42; H. Diekmann, M. Walger, and H. von Wedel, "Die Riechleistungen von Gehorlosen und Blinden [Sense of smell in deaf and blind patients]," *HNO* 42 (1994):264–69; O. Schwenn, I. Hundorf, et al., "Können Blinde besser riechen als Normalsichtige? [Do blind persons have a better sense of smell than normal sighted people?]," *Klinische Monatsblätter für Augenheilkunde* 219 (2002):649–54.

57   *in three of the six studies*   Blind are better: Murphy & Cain 1986; Wakefield et al., 2004; Rosenbluth et al., 2000; blind no better: Diekmann et al., 1994; Schwenn et al., 2002; Smith et al., 1993.

58   *the master's view*   A. A. Brill, "The sense of smell in the neuroses and psychoses," *The Psychoanalytic Quarterly* 1 (1932):17–42.

58   *The original texts*   On November 14, 1897, Freud wrote to his colleague Wilhelm Fliess in Berlin and speculated about a biological basis for the psychological repression of sexual impulses; see *The Complete Letters of Sigmund Freud to Wilhelm Fliess 1887–1904*, edited by J. M. Masson (Belknap/Harvard University Press, 1985), pp. 278–82.

58   *"audacious, highly speculative"*   Sigmund Freud, *Civilization and Its Discontents*, translated by James Strachey with introduction by Peter Gay (New York: W. W. Norton, 1989).

59   *helped devalue smell*   Annick Le Guérer, "Olfaction and cognition: A philosophical and psychoanalytic overview," in C. Rouby, B. Schaal, et al., eds., *Olfaction, Taste, and Cognition* (Cambridge: Cambridge University Press, 2002), p. 6.

59   *University of Texas study*   D. Singh and P. M. Bronstad, "Female body odour is a potential cue to ovulation," *Proceedings of the Royal Society of London, Series B, Biological Sciences* 268 (2001):797–801.

59   *psychologist Paul Rozin*   A. Wrzesniewski, C. McCauley, and P. Rozin, "Odor and affect: Individual differences in the impact of odor on liking for places, things and people," *Chemical Senses* 24 (1999):713–21.

59   *psychoanalyst Annick Le Guérer*   Le Guérer, in C. Rouby, B. Schaal, et al., eds., 2002, p. 6.

59   *anthropologist David Howes*   David Howes, "Freud's nose: The repression of nasality and the origin of psychoanalytic theory," in Victoria De Rijke, Lene Østermark-Johansen, and Helen Thomas, eds., *Nose Book: Representa-*

tions of the Nose in Literature and the Arts (London: Middlesex University Press, 2000), pp. 265–81.

60 *a medical disaster zone*   Frank J. Sulloway, *Freud, Biologist of the Mind* (New York: Basic Books, 1979), p. 143. See also Max Schur, *Freud: Living and Dying* (International Universities Press, 1972), pp. 77–90 for details of operation; Ernest Jones, *The Life and Work of Sigmund Freud*, vol. 1, *1856–1900: The Formative Years and the Great Discoveries* (New York: Basic Books, 1953), pp. 308–9.

60 *"No doubt there is a vast difference"*   W. H. Hudson, "On the Sense of Smell," *The Century Magazine*, August 1922, pp. 497–506.

61 *a man-bites-dog story*   A. N. Gilbert, K. Yamazaki, et al., "Olfactory discrimination of mouse strains (*Mus musculus*) and major histocompatibility types by humans (*Homo sapiens*)," *Journal of Comparative Psychology* 100 (1986):262–65.

61 *impressive man-smells-dog story*   D. L. Wells and P. G. Hepper, "The discrimination of dog odours by humans," *Perception* 29 (2000):111–15.

62 *dogs can sniff out bladder cancer*   C. M. Willis, S. M. Church, et al., "Olfactory detection of human bladder cancer by dogs: Proof of principle study," *BMJ* 329 (2004):712–14; but also see M. Leahy, "Olfactory detection of human bladder cancer by dogs: Cause or association?" and J. S. Welsh, "Olfactory detection of human bladder cancer by dogs: Another cancer detected by 'pet scan,' " in ibid., 1286–87.

62 *Just by smelling some ice cream*   S. Jiamyangyuen, J. F. Delwiche, and W. J. Harper, "The impact of wood ice cream sticks' origin on the aroma of exposed ice cream mixes," *Journal of Dairy Science* 85 (2002):355–59.

63 *Feynman had a great party trick*   Richard P. Feynman, *"Surely You're Joking, Mr. Feynman!": Adventures of a Curious Character* (New York: W. W. Norton, 1985).

63 *hand odor is individually distinctive*   P. Wallace, "Individual discrimination of humans by odor," *Physiology & Behavior* 19 (1977):577–79.

63 *the quintessential doggy task*   J. Porter, B. Craven, et al., "Mechanisms of scent-tracking in humans," *Nature Neuroscience* 10 (2007):27–29.

63 *drug dogs and humans have almost identical sensitivity*   Lorenzo, Wan, et al., "Laboratory and field experiments," p. 1213.

63 *"of extremely slight service"*   Charles Darwin, *The Descent of Man and Selection in Relation to Sex*, vol. 1 (London: John Murray, 1871), pp. 23–24.

64   *"Among the apes it has greatly lost importance"*   Havelock Ellis, *Studies in the Psychology of Sex: Sexual Selection in Man* (Philadelphia: F. A. Davis Company, 1922), pp. 47, 48.

64   *"The sense of smell in primates is greatly reduced"*   S. Rouquier, A. Blancher, and D. Giorgi, "The olfactory receptor gene repertoire in primates and mouse: Evidence for reduction of the functional fraction in primates," *Proceedings of the National Academy of Sciences USA*, 97 (2000):2870–74.

64   *questioning the textbook distinction*   T. D. Smith, K. P. Bhatnagar, et al., "Distribution of olfactory epithelium in the primate nasal cavity: Are microsmia and macrosmia valid morphological concepts?" *Anatomical Record* 281 (2004):1173–81; T. D. Smith and K. P. Bhatnagar, "Microsmatic primates: Reconsidering how and when size matters," *Anatomical Record* 279 (2004):24–31.

64   *neurobiologist Gordon Shepherd*   G. M. Shepherd, "The human sense of smell: Are we better than we think?" *PLoS Biology* 2 (2004):572–75.

64   *sensory physiologist Mathias Laska*   M. Laska, D. Genzel, and A. Wieser, "The number of functional olfactory receptor genes and the relative size of olfactory brain structures are poor predictors of olfactory discrimination performance with enantiomers," *Chemical Senses* 30 (2005):171–75; M. Laska, A. Wieser, et al., "Olfactory responsiveness to two odorous steroids in three species of nonhuman primates," *Chemical Senses* 30 (2005):505–11.

65   *New evidence suggests*   P. Quignon, E. Kirkness, et al., "Comparison of the canine and human olfactory receptor gene repertoires," *Genome Biology* 4 (2003):R80; Y. Gilad, O. Man, and G. Glusman, "A comparison of the human and chimpanzee olfactory receptor gene repertoires," *Genome Research* 15 (2005):224–30; Quignon, et al., "The dog and rat olfactory receptor repertoires," *Genome Biology* 6 (2005):R83, pp. 1–9.

65   *compare odor receptor subfamilies*   P. Quignon, et al., "The dog and rat olfactory receptor repertoires," *Genome Biology* 6 (2005):R83, pp. 1–9; P. A. Godfrey, B. Malnic, and L. B. Buck, "The mouse olfactory receptor gene family," *Proceedings of the National Academy of Sciences USA* 101 (2004):2156–61; B. Malnic, P. A. Godfrey, and L. B. Buck, "The human olfactory receptor gene family," *Proceedings of the National Academy of Sciences USA* 101 (2004):2584–89; Y. Gilad, O. Man, and G. Glusman, "A comparison of the human and chimpanzee olfactory receptor gene repertoires," *Genome Research* 15 (2005):224–30.

66   *"One morning when I walked"*   Jack Kornfield, *A Path with Heart: A Guide Through the Perils and Promises of Spiritual Life* (New York: Bantam, 1993), p. 125.

67   *Experts outperform novices*   M. Bende and S. Nordin, "Perceptual learning in olfaction: Professional wine tasters versus controls," *Physiology & Behavior* 62 (1997):1065–70; H. T. Lawless, "Flavor description of white wine by 'expert' and nonexpert wine consumers," *Journal of Food Science* 49 (1984):120–23; W. V. Parr, D. Heatherbell, and K. G. White, "Demystifying wine expertise: Olfactory threshold, perceptual skill and semantic memory in expert and novice wine judges," *Chemical Senses* 27 (2002):747–55.

67   *their job can be done with only an adequate nose*   Calkin and Jellinek, *Perfumery,* p. 3.

67   *better olfactory imagery ability*   A. N. Gilbert, M. Crouch, and S. E. Kemp, "Olfactory and visual mental imagery," *Journal of Mental Imagery* 22 (1998):137–46.

67   *professional perfume researchers*   Byung-Chan Min, et al., "Analysis of mutual information content for EEG responses to odor stimulation for subjects classified by occupation," *Chemical Senses* 28 (2003):741–49.

68   *brain activity in wine sommeliers*   A. Castriota-Scanderbeg, G. E. Hagberg, et al., "The appreciation of wine by sommeliers: A functional magnetic resonance study of sensory integration," *Neuroimage* 25 (2005): 570–78.

68   *Süskind's novel*   Patrick Süskind, *Perfume: The Story of a Murderer,* translated by John E. Woods (New York: Alfred A. Knopf, 1986).

69   *"the fragrances poured into me"*   Salman Rushdie, *Midnight's Children* (New York: Alfred A. Knopf, 1980), p. 378.

70   *"I suggest that if the police"*   Helen Keller, *Midstream: My Later Life* (New York: Crowell Publishing Co., 1929), p. 165.

70   *the bills reeked of marijuana*   Associated Press, "Smelly Money Lands Indiana Man in Jail," April 7, 2005.

71   *made its way to the Ohio Supreme Court*   State v. Moore (2000), 90 Ohio St.3d 47.

72   *forensic sniff tests*   R. L. Doty, T. Wudarski, et al., "Marijuana odor perception: Studies modeled from probable cause cases," *Law and Human Behavior* 28 (2004):223–33.

72   *detecting drunk drivers*   H. Moskowitz, M. Burns, and S. Ferguson, "Police officers' detection of breath odors from alcohol ingestion," *Accident Analysis and Prevention* 31 (1999):175–80.

73   *no corroborating evidence is needed*   E. Hendrie, "The motor vehicle exception," *FBI Law Enforcement Bulletin* 74, no. 8 (August 2005).

73   *cited by the defense*   United States of America v. Burton Dean Viers, CA No. 06-30266, Appellant's Opening Brief.

Chapter 4. The Art of the Sniff

75   *Early experiments were ingenious*   Edwin G. Boring, *Sensation and Perception in the History of Experimental Psychology* (New York: D. Appleton-Century Co., 1942), p. 440.

75   *A second, more grotesque experiment*   E. Paulsen, "Experimentelle Untersuchungen über die Strömung der Luft der Nasenhöhle," *Sitzungber. d. preuss. Akad. d. Wiss.* 85 (1882):328.

76   *sophisticated computer models*   K. Zhao, P. Dalton, et al., "Numerical modeling of turbulent and laminar airflow and odorant transport during sniffing in the human and rat nose," *Chemical Senses* 31 (2006):107–18.

76   *a highly regarded neurological surgeon*   J. S. Oppenheim, "Neurosurgery at the Mount Sinai Hospital," *Journal of Neurosurgery* 80 (1994):935–38.

76   *came up with a method*   C. A. Elsberg and I. Levy, "The sense of smell (I): A new and simple method of quantitative olfactometry," *Bulletin of the Neurological Institute of New York* 4 (1935):5–19; Elsberg, Levy, and E. D. Brewer, "A new method for testing the sense of smell and for the establishment of olfactory values for odorous substance," *Science* 83 (1936):211–12.

77   *Zwaardemaker's device*   H. Zwaardemaker, "Präzisionsolfaktometrie." *Arch. für Layng. und Rhinol.* 15 (1904):171–77.

77   *Elsberg's results*   *Time*, November 25, 1935, p. 40; *New York Times*, November 13, 1935.

78   *psychology professor at UCLA*   F. Nowell Jones; "A test of the validity of the Elsberg method of olfactometry," *American Journal of Psychology* 66 (1953):81–85; "The reliability of olfactory thresholds obtained by sniffing," *American Journal of Psychology* 68 (1955):289–90; "A comparison of the methods of olfactory stimulation: Blasting vs. sniffing," ibid., 486–88. He pulls his punches somewhat in the later paper, but the deed was done.

78   *"we might be better off today"*   B. M. Wenzel, "Problems of odor research from the viewpoint of a psychologist," *Annals of the New York Academy of Sciences* 58 (1954):58–61.

78   *beginning in 1982*   D. G. Laing, "Characterisation of human behaviour during odour perception," *Perception* 11 (1982):221–30.

80   *"a single natural sniff"*   D. G. Laing, "Natural sniffing gives optimum odour

perception for humans," *Perception* 12 (1983):99–117; "Identification of single dissimilar odors is achieved by humans with a single sniff," *Physiology & Behavior* 37 (1986):163–70; "Optimum perception of odor intensity by humans," *Physiology & Behavior* 34 (1985):569–74.

80   *The dictionary's dichotomy*   *Oxford English Dictionary*, 2nd Edition, 1989.

80   *Berkeley smell researcher Noam Sobel*   N. Sobel, V. Prabhakaran, et al., "Odorant-induced and sniff-induced activation in the cerebellum of the human," *Journal of Neuroscience* 18 (1998):8990–9001; B. N. Johnson, J. D. Mainland, and N. Sobel, "Rapid olfactory processing implicates subcortical control of an olfactomotor system," *Journal of Neurophysiology* 90 (2003):1084–94; N. Sobel, V. Prabhakaran, et al., "Sniffing and smelling: Separate subsystems in the human olfactory cortex," *Nature* 392 (1998): 282–86.

81   *imagined odors*   M. Bensafi, J. Porter, et al., "Olfactomotor activity during imagery mimics that during perception," *Nature Neuroscience* 6 (2003): 1142–44.

81   *"the sniff is part of the percept"*   J. Mainland and N. Sobel, "The sniff is part of olfactory percept," *Chemical Senses* 31 (2006):181–96.

82   *A new smell test*   R. A. Frank, M. F. Dulay, et al., "A comparison of the sniff magnitude test and the University of Pennsylvania Smell Identification Test in children and nonnative English speakers," *Physiology & Behavior* 81 (2004):475–80; Frank, Dulay, and R. C. Gesteland, "Assessment of the Sniff Magnitude Test as a clinical test of olfactory function," *Physiology & Behavior* 78 (2003):195–204; Frank, Gesteland, et al., "Characterization of the sniff magnitude test," *Archives of Otolaryngology—Head & Neck Surgery* 132 (2006):532–36.

82   *"We have glasses to help"*   "Can a Robot Have a Nose?" *Popular Science Monthly*, October 1931, p. 70.

83   *resembles a polite yawn*   B. Risberg-Berlin, R. Ylitalo, and C. Finizia, "Screening and rehabilitation of olfaction after total laryngectomy in Swedish patients: Results from an intervention study using the Nasal Airflow-Inducing Maneuver," *Archives of Otolaryngology—Head & Neck Surgery* 132 (2006):301–6.

83   *tracheostomy valve*   S. W. Lichtman, I. L. Birnbaum, et al., "Effect of a tracheostomy speaking valve on secretions, arterial oxygenation, and olfaction: A quantitative evaluation," *Journal of Speech and Hearing Research* 38 (1995):549–55; D. S. Braz, M. M. Ribas, et al., "Quality of life and depression in patients undergoing total and partial laryngectomy," *Clinics* 60 (2005):135–42.

83 *Parkinson's disease*   N. Sobel, M. E. Thomason, et al., "An impairment in sniffing contributes to the olfactory impairment in Parkinson's disease," *Proceedings of the National Academy of Sciences USA* 98 (2001):4154–59.

83 *device to help the sniff-impaired*   Roy F. Knight, "Smelling Aid Device," U.S. Patent 5,522,253, issued June 4, 1996.

83 *first marketed in 1993*   "GlaxoSmithKline to Acquire Nasal-Strip Maker CNS," *Wall Street Journal*, October 10, 2006.

84 *dilator makes odors smell stronger*   D. E. Hornung, C. Chin, et al., "Effect of nasal dilators on perceived odor intensity," *Chemical Senses* 22 (1997):177–80; D. E. Hornung, D. J. Smith, et al., "Effect of nasal dilators on nasal structures, sniffing strategies, and olfactory ability," *Rhinology* 39 (2001):84–87.

84 *intensity of food aromas*   B. Raudenbush and B. Meyer, "Effect of nasal dilators on pleasantness, intensity and sampling behaviors of foods in the oral cavity," *Rhinology* 39 (2001):80–83.

85 *Even one minute*   P. Dalton and C. J. Wysocki, "The nature and duration of adaptation following long-term odor exposure," *Perception & Psychophysics* 58 (1996):781–92.

87 *"While awaiting results"*   E. E. Slosson, "A lecture experiment in hallucinations," *Psychological Review* 6 (1899):407–8. A few years later, A. S. Edwards at Cornell University got similar results in the lab, published as "An experimental study of sensory suggestion," *American Journal of Psychology* 26 (1915):99–129.

88 *sensory expert Michael O'Mahony*   M. O'Mahony, "Smell illusions and suggestion: Reports of smells contingent on tones played on television and radio," *Chemical Senses & Flavor* 3 (1978):183–89.

88 *We sprayed water mist*   S. C. Knasko, A. N. Gilbert, and J. Sabini, "Emotional state, physical well-being and performance in the presence of feigned ambient odor," *Journal of Applied Social Psychology* 20 (1990):1345–57.

89 *sit in a test chamber*   For references to Dalton's work, see P. Dalton, "Cognitive influences on health symptoms from acute chemical exposure," *Health Psychology* 18 (1999):579–90.

89 *an authority figure in a lab coat*   P. Dalton, "Odor, irritation and perception of health risk," *International Archives of Occupational and Environmental Health* 75 (2002):283–90.

90 *reverse the aromatherapy effects*   C. E. Campenni, E. J. Crawley, and M. E.

Meier, "Role of suggestion in odor-induced mood change," *Psychological Reports* 94 (2004):1127–36.

90 *Norwegian air ambulance* "Near disaster after warning stink ignored," *Aftenposten*, June 5, 2003.

Chapter 5. A Nose for the Mouth

91 *"Blindfold a person"* H. T. Finck, "The gastronomic value of odours," *Contemporary Review* 50 (1886):680–95.

91 *The discovery in 1996* N. Chaudhari, H. Yang, et al., "The taste of monosodium glutamate: Membrane receptors in taste buds," *Journal of Neuroscience* 16 (1996):3817–26.

92 *those who claim* Carl Sagan, *The Dragons of Eden* (New York: Random House, 1977), p. 156; Andrew Hamilton, "What science is learning about smell," *Science Digest*, November 1966, pp. 81–84; Havelock Ellis, *Studies in the Psychology of Sex: Sexual Selection in Man* (Philadelphia: F. A. Davis Company, 1922), pp. 47, 48.

93 *psychologist Paul Rozin* P. Rozin, " 'Taste-smell confusions' and the duality of the olfactory sense," *Perception & Psychophysics* 31 (1982):397–401.

93 *psychologist Debra Zellner* B. J. Koza, A. Cilmi, et al., "Color enhances orthonasal olfactory intensity and reduces retronasal olfactory intensity," *Chemical Senses* 30 (2005):643–49.

94 *Australian psychologist R. J. Stevenson* Stevenson's work is described in D. M. Small and J. Prescott, "Odor/taste integration and the perception of flavor," *Experimental Brain Research* 166 (2005):345–57.

94 *in the other direction as well* J. Djordjevic, R. J. Zatorre, and M. Jones-Gotman, "Odor-induced changes in taste perception," *Experimental Brain Research* 159 (2004):405–8.

95 *"no human populations"* R. Wrangham and N. Conklin-Brittain, "Cooking as a biological trait," *Comparative biochemistry and physiology, Part A, Molecular & Integrative Physiology* 136 (2003):35–46. See also A. Gibbons, "Paleoanthropology: Food for thought," *Science* 316 (2007):1558–60.

96 *more nimble modern mouth* P. W. Lucas, K. Y. Ang, et al., "A brief review of the recent evolution of the human mouth in physiological and nutritional contexts," *Physiology & Behavior* 89 (2006):36–38.

96 *The aroma of bacon* M. R. Yeomans, "Olfactory influences on appetite and satiety in humans," *Physiology & Behavior* 87 (2006):800–804.

97 *"any dried, fragrant, aromatic"* Kenneth T. Farrell, *Spices, Condiments, and Seasonings*, 2nd edition (New York: Springer, 1990).

97 *"flavor principles"* Elisabeth Rozin, *Ethnic Cuisine: How to Create the Authentic Flavors of 30 International Cuisines* (New York: Penguin, 1992), p. xiv.

98 *"You can prepare forty dishes"* Michael Washburn, "Q&A: Chewing the Fat with Charlie Trotter," *Detours: The Online Magazine of the Illinois Humanities Council* 5, no. 1 (May 2003).

99 *To test their idea* J. Billing and P. W. Sherman, "Antimicrobial functions of spices: Why some like it hot," *Quarterly Review of Biology* 73 (1998):3–49.

100 *vegetable dishes* P. W. Sherman and G. A. Hash, "Why vegetable recipes are not very spicy," *Evolution and Human Behavior* 22 (2001):147–63.

101 *"the price of civilization"* Nicholas Wade, *Before the Dawn* (New York: Penguin, 2006), p. 270.

101 *Olfaction is one such hot spot* B. F. Voight, S. Kudaravalli, et al., "A map of recent positive selection in the human genome," *PLoS Biology* 4 (2006):446; R. Nielsen, C. Bustamante, et al., "A scan for positively selected genes in the genomes of humans and chimpanzees," *PLoS Biology* 3 (2005):e170; A. G. Clark, S. Glanowski, et al., "Inferring nonneutral evolution from human-chimp-mouse orthologous gene trios," *Science* 302 (2003):1960–63; S. H. Williamson, M. J. Hubisz, et al., "Localizing recent adaptive evolution in the human genome," *PLoS Genetics* 3 (2007):e90; H. Tang, S. Choudhry, et al., "Recent genetic selection in the ancestral admixture of Puerto Ricans," *American Journal of Human Genetics* 81 (2007):626–33.

101 *gene for lactose absorption* T. Bersaglieri, P. C. Sabeti, et al., "Genetic signatures of strong recent positive selection at the lactase gene," *American Journal of Human Genetics* 74 (2004):1111–20; S. A. Tishkoff, F. A. Reed, et al., "Convergent adaptation of human lactase persistence in Africa and Europe," *Nature Genetics* 39 (2007):31–40; J. Burger, M. Kirchner, et al., "Absence of the lactase-persistence-associated allele in early Neolithic Europeans," *Proceedings of the National Academy of Sciences USA* 104 (2007):3736–41.

102 *"has delivered a richer repertoire"* Shepherd, "The human sense of smell," p. 573.

103 *Smell prejudice* Pearl Buck, *The Good Earth* (New York: John Dan, 1931), pp. 110–11.

103 *olfactory stereotyping* Constance Classen, David Howes, and Anthony Synnott, *Aroma: The Cultural History of Smell* (London: Routledge, 1994).

103 *Scottish clans* Ron Clark, "Savouring the Sweet Smell of Scotland: Aroma

Scientist Makes Scents of the Natural World," *The Herald* (Glasgow), March 14, 2003.

104   *At cultural boundaries*   S. Ayabe-Kanamura, I. Schicker, et al., "Differences in perception of everyday odors: A Japanese-German cross-cultural study," *Chemical Senses* 23 (1998):31–38.

104   *Jha's novel*   Radhika Jha, *Smell: A Novel* (New York: SoHo Press, 1999).

105   *"a repulsive gelatinous fishlike dish"*   Garrison Keillor, *Lake Wobegon Days* (New York: Viking, 1985).

105   *Brown compiled a list*   For Brown's list, see Steven Pinker, *The Blank Slate* (New York: Viking, 2002).

105   *"a miasma of eyeglass-fogging kimchi breath"*   P. J. O'Rourke, *Holidays in Hell* (New York: Grove/Atlantic, 2000), p. 46.

106   *Mango Chipotle seafood marinade*   Tom Van Riper, "Turning Up the Heat," Forbes.com, March 21, 2006.

106   *"West Coast doughnut flour"*   Francis Sill Wickware, "They're After Your Nose Now," *The Saturday Evening Post*, June 21, 1947, p. 26.

106   *air-freshener sales*   "Trends in Air Care," data presented by Lynn Dornblaser, GNPD Consulting Services/Mintel Group, 2005.

106   *beer is less bitter*   Sarah Ellison, "After Making Beer Ever Lighter, Anheuser Faces a New Palate," *Wall Street Journal*, April 26, 2006.

107   *tasting more and more alike*   "The French Move Their Cheese—Down-Market," *Wall Street Journal*, June 20, 2000; "U.S., France Clash over Curdled Milk; Defending France's Smelliest Cheese," *Wall Street Journal*, May 27, 1999; "Sweet Stink of Success," *The Guardian* (London), November 26, 1999.

107   *"Joel Lloyd Bellenson places a little ceramic bowl"*   Charles Platt, "You've Got Smell!" *Wired*, issue 7.11 (November 1999), p. 256.

108   *aroma impact molecules*   M. Czerny, F. Mayer, and W. Grosch, "Sensory study on the character impact odorants of roasted arabica coffee," *Journal of Agricultural and Food Chemistry* 47 (1999):695–699.

108   *"refresh the nose"*   Crocker and Henderson (1927).

109   *the results were surprising*   E. A. Johnson and Z. M. Vickers, "The effectiveness of palate cleansers," Presentation to the Institute of Food Technologists annual meeting 2002.

109 *sensory specialist Hildegaarde Heymann*   B. Madrigal-Galan and H. Heymann, "Sensory effects of consuming cheese prior to evaluating red wine flavor," *American Journal of Enology and Viticulture* 57 (2006):12–22. Also see, "UC Davis Study Challenges Classic Wine-Cheese Pairings," *San Francisco Chronicle,* July 16, 2005.

110ʹ *Only three studies*   J. F. Delwiche and M. L. Pelchat, "Influence of glass shape on wine aroma," *Journal of Sensory Studies* 17 (2002):19–28; J. F. Delwiche, "The impact of glass shape on the perception of wine: Bacchus to the future," Proceedings of the Inaugural Brock University Wine Conference, 2002; T. Hummel, J. F. Delwiche, et al., "Effects of the form of glasses on the perception of wine flavors: A study in untrained subjects," *Appetite* 41 (2003):197–202.

Chapter 6. The Malevolence of Malodor

111 *"And when euyl substance"*   John Reidy, ed., *Thomas Norton's Ordinal of Alchemy,* Early English Text Society no. 272 (Oxford University Press, 1975), p. 64.

112 *"some stinking doonghills"*   Danielle Nagler, "Towards the smell of mortality: Shakespeare and the ideas of smell 1588–1625," *The Cambridge Quarterly* 26 (1997):42–58.

112 *"Ten years from now"*   James Bovard, "Get a Whiff of This!", *Wall Street Journal,* December 27, 1995.

113 *"a poorly understood and controversial syndrome"*   O. Van den Bergh, K. Stegen, et al., "Acquisition and extinction of somatic symptoms in response to odours: A Pavlovian paradigm relevant to multiple chemical sensitivity," *Occupational and Environmental Medicine* 56 (1999):295–301.

113 *results consistently show*   R. L. Doty, D. A. Deems, et al., "Olfactory sensitivity, nasal resistance, and autonomic function in patients with multiple chemical sensitivities," *Archives of Otolaryngology—Head & Neck Surgery* 114 (1988):1422–27; E. Caccappolo, H. Kipen, et al., "Odor perception: Multiple chemical sensitivities, chronic fatigue, and asthma," *Journal of Occupational and Environmental Medicine* 42 (2000):629–38; D. Papo, B. Eberlein-Konig, et al., "Chemosensory function and psychological profile in patients with multiple chemical sensitivity: Comparison with odor-sensitive and asymptomatic controls," *Journal of Psychosomatic Research* 60 (2006):199–209.

114 *In another test*   Caccappolo, Kipen, et al., "Odor perception," pp. 629–38.

114 *the smell of baby oil*   Izabella St. James, *Bunny Tales: Behind Closed Doors at the Playboy Mansion* (Philadelphia: Running Press, 2006).

115   *Galveston hurricane*   Herbert Molloy Mason Jr., *Death from the Sea* (Dial Press, 1972), pp. 198–99.

115   *funeral home scandal*   "A Mortuary Tangled in the Macabre," *Los Angeles Times,* December 30, 1988; Kathy Braidhill, *Chop Shop* (New York: Pinnacle Books, 1993,) p. 138.

115   *fire department paramedic*   E. Vermetten and J. D. Bremner, "Olfaction as a traumatic reminder in posttraumatic stress disorder: Case reports and review," *Journal of Clinical Psychiatry* 64 (2003):202–7.

116   *Cambodian refugees*   D. Hinton, V. Pich, et al., "Olfactory-triggered panic attacks among Khmer refugees: A contextual approach," *Transcultural Psychiatry* 41 (2004):155–99; D. E. Hinton, V. Pich, et al., "Olfactory-triggered panic attacks among Cambodian refugees attending a psychiatric clinic," *General Hospital Psychiatry* 26 (2004):390–97.

116   *makes people sick*   W. Winters, S. Devriese, et al., "Media warnings about environmental pollution facilitate the acquisition of symptoms in response to chemical substances," *Psychosomatic Medicine* 65 (2003):332–38.

117   *single episode*   O. Van den Bergh, P. J. Kempynck, et al., "Respiratory learning and somatic complaints: A conditioning approach using CO_2-enriched air inhalation, *Behaviour Research and Therapy* 33 (1995):517–27.

117   *stimulus generalization*   S. Devriese, W. Winters, et al., "Generalization of acquired somatic symptoms in response to odors: A pavlovian perspective on multiple chemical sensitivity, *Psychosomatic Medicine* 62 (2000):751–59.

117   *phenomenon called extinction*   O. Van den Bergh, K. Stegen, et al., Acquisition and extinction of somatic symptoms in response to odours: A Pavlovian paradigm relevant to multiple chemical sensitivity, *Occupational and Environmental Medicine* 56 (1999):295–301.

118   *"warnings and campaigns"*   W. Winters, S. Devriese, et al., "Media warnings about environmental pollution facilitate the acquisition of symptoms in response to chemical substances," *Psychosomatic Medicine* 65 (2003):332–38.

118   *If you believe*   S. Devriese, W. Winters, et al., "Perceived relation between odors and a negative event determines learning of symptoms in response to chemicals," *International Archives of Occupational and Environmental Health* 77 (2004):200–204.

118   *"hypothesized biological processes"*   H. Staudenmayer, K. E. Binkley, et al., "Idiopathic environmental intolerance, Part 1: A causation analysis applying Bradford Hill's criteria to the toxicogenic theory," *Toxicological Reviews* 22 (2003):235–46. A meta-analysis that casts doubt on the validity of MCS as

a clinical construct and finds "expectations and prior beliefs" to be a key factor in response is to be found in J. Das-Munshi, G. J. Rubin, and S. Wessely, "Multiple chemical sensitivities: A systematic review of provocation studies," *Journal of Allergy and Clinical Immunology* 118 (2006):1257–64.

118   *a psychogenic theory*   H. Staudenmayer, K. E. Binkley, et al., "Idiopathic environmental intolerance, Part 2: A causation analysis applying Bradford Hill's criteria to the psychogenic theory," *Toxicological Reviews* 22 (2003): 247–61. See also D. Papo, B. Eberlein-Konig, et al., "Chemosensory function and psychological profile in patients with multiple chemical sensitivity: Comparison with odor-sensitive and asymptomatic controls," *Journal of Psychosomatic Research* 60 (2006):199–209.

119   *"Imagination has, besides, a great deal"*   Eugene Rimmel, *The Book of Perfumes*, 7th edition (London: Chapman and Hall, 1871), p. 13.

120   *"Now, can it be possible"*   Mark Twain, "About Smells," *The Galaxy*, May 1870.

121   *"battle royal"*   "Ex-con, Woman Dead in Bronx," *New York Daily News*, September 10, 2004.

121   *can happen anywhere*   "Fears of Growing Old Lead Couple to Suicide," *Chicago Sun-Times*, August 12, 1994; "Pair in 80s Found Dead in Home; Apparently Died Several Weeks Ago," *Houston Chronicle*, March 22, 1997.

121   *O. J. Simpson*   "What's Become of O.J.'s Ex-Gal Pal?" *New York Daily News*, January 23, 2002; "Miami Cops Say O.J.'s Ex-Girlfriend Isn't Missing, Knows About Dead Cat," *South Florida Sun-Sentinel*, January 23, 2003.

122   *urban legend*   Jan Harold Brunvand, *The Baby Train and Other Lusty Urban Legends* (New York: Norton, 1993); Barbara Mikkelson writing on Snopes.com.

122   *hit man immortalized*   Katherine Ramsland, "Richard Kuklinski: The Iceman," chapter title "Going to Florida," Crimelibrary.com.

122   *Jerry Payne*   Jessica Snyder Sachs, *Corpse: Nature, Forensics, and the Struggle to Pinpoint Time of Death* (Cambridge, MA: Perseus Books, 2001).

123   *an ironic twist*   "Woman, two men dead in Bronx apt. bloodbath," *New York Daily News*, January 22, 2002.

123   *found in bad breath*   S. Goldberg, A. Kozlovsky, et al., "Cadaverine as a putative component of oral malodor," *Journal of Dental Research* 73 (1994):1168–72.

124   *"smelled like kim chee"*   "Isle Mainland Traveler Shared Room with Corpse," *Honolulu Star-Bulletin,* August 1, 1996.

124   *lives at the scene of the crime*   "He Slays Wife, Then Can't Take Smell, Say Cops," *New York Daily News,* December 11, 2003; "Mom, Stepdad Charged in Death of Disabled Man," *Houston Chronicle,* April 11, 2002.

124   *made of tougher stuff*   "Body Undiscovered in Apartment for 2 Years?" *Tucson Citizen,* April 8, 2005; "Woman Drove for Days with Dead Mother," Reuters/CNN, April 29, 2004.

125   *young hiker*   Aron Ralston, *Between a Rock and a Hard Place* (New York: Atria Books, 2004).

Chapter 7. The Olfactory Imagination

128   *"The rank effluvium"*   Charles Darwin, *The Descent of Man, and Selection in Relation to Sex* (London: John Murray, 1871), p. 279.

129   *"with a confidence that always astonished"*   Edouard Toulouse, *Enquête médico-psychologique sur les rapports de la supériorité intellectuale avec le névrophatie* (Paris Societé d'edition scientifiques, 1896); *Annales Médico-Psychologiques,* series 8, vol. 5 (1897):425–46.

130   *consumer survey*   Sarah Purcell, "Scents and Scentsibilities," *Chemist & Druggist,* November 22, 2003, p. S32.

130   *Cobain's personal journals*   Tim Appelo, "Kurt Cobain's Last No. 1 Hit," *Seattle Weekly,* December 25, 2002; Kurt Cobain, *Journals* (Riverhead Books, 2002).

130   *His favorite book*   Charles R. Cross, *Heavier than Heaven: A Biography of Kurt Cobain* (New York: Hyperion, 2001).

131   *world's first Aroma Jockey*   E-mail correspondence with Eric Berghammer, aka Odo7, May–June 2005.

131   *With my colleagues*   A. N. Gilbert, M. Crouch, and S. E. Kemp, "Olfactory and visual mental imagery," *Journal of Mental Imagery* 22 (1998): 137–46.

132   *Other researchers have used our test*   M. Bensafi and C. Rouby, "Individual differences in odor imaging ability reflect differences in olfactory and emotional perception," *Chemical Senses* 32 (2007):237–44.

132   *innovative American director*   Lise-Lone Marker, *David Belasco: Naturalism in the American Theatre* (Princeton, NJ: Princeton University Press, 1975), p. 61 ff.

133 *pioneers of olfactory multimedia*   Beatriz Colomina, "Enclosed by images: The Eameses' multimedia architecture, *Grey Room* 2, (Winter 2001), pp. 6–29 (esp. p. 13); Stanley Abercrombie, *George Nelson: The Design of Modern Design* (Boston: MIT Press, 1995), p 147; Colomina, 2001, p. 14, referencing Eames collaborator George Nelson's quote in Abercrombie, 1995; Owen Gingerich, "A conversation with Charles Eames," *The American Scholar* 46, no. 3 (1977):326–37 (esp. p. 331); Abercrombie, *George Nelson*, p. 147.

134 *written descriptions evoke*   J. C. Baird and K. A. Harder, "The psychophysics of imagery," *Perception & Psychophysics* 62 (2000):113–26; J. Gonzalez, A. Barros-Loscertales, et al., "Reading cinnamon activates olfactory brain regions," *Neuroimage* 32 (2006):906–12.

134 *In a letter*   Helen McAfee, "The Sense of Smell," *The Nation*, January 15, 1914, pp. 57–58.

134 *"This all started"*   Anne Tyler, *Ladder of Years* (New York: Alfred A. Knopf, 1995).

135 *"You get down on your knees"*   Jay McInerney, *Bright Lights, Big City* (New York: Vintage, 1984).

135 *smell in cocaine abusers*   A. S. Gordon, D. T. Moran, et al., "The effect of chronic cocaine abuse on human olfaction," *Archives of Otolaryngology—Head & Neck Surgery* 116 (1990):1415–18.

135 *"The chimney of the new house"*   *Nathaniel Hawthorne: Collected Novels* (New York: Library of America edition, 1983), p. 360.

136 *the best smell-based story in American letters*   *Nathaniel Hawthorne: Tales and Sketches* (New York: Library of America, 1982).

137 *lived her entire life*   Judith Farr, *The Gardens of Emily Dickinson* (Cambridge, MA: Harvard University Press, 2004).

138 *"cultivation of emotional intensity"*   Agnieszka Salska, "Dickinson's letters: From correspondence to poetry," in *The Emily Dickinson Handbook*, edited by G. Grabher, R. Hagenbüchle, and C. Miller (Amherst: University of Massachusetts Press, 1998).

138 *Paglia challenged this admiring consensus*   Camille Paglia, *Sexual Personae: Art and Decadence from Nefertiti to Emily Dickinson* (New Haven: Yale University Press, 1990).

138 *In her poems*   Poems quoted here can be found as numbered in *The Complete Poems of Emily Dickinson*, edited by T. H. Johnson (Boston: Little,

Brown, 1960): drinker #1628, Inebriate #214, quaffing #230, kill your balm #238, when it dies #333, little odor #785, oils are wrung #675.

139 *"fetishistic fascination"* Marc A. Weiner, "Wagner's Nose and the Ideology of Perception." *Monatshefte* 81 (1989):62–78.

140 *"The night is cool"* Leopold von Sacher-Masoch, *Venus in Furs* (1870; translation by Fernanda Savage, 1921).

141 *American novelist Willa Cather* Marilee Lindemann, *Willa Cather: Queering America* (New York: Columbia University Press, 1999); Willa Cather, *O Pioneers!* (Boston/New York: Houghton Mifflin Company, 1913).

142 *"maybe smell is one of my"* Frederick L. Gwynn and Joseph L. Blotner, *Faulkner in the University* (Charlottesville: University of Virginia Press, 1959), p. 253.

142 *it doesn't add up* Joseph Blotner, *Faulkner: A Biography* (New York: Random House, 1974).

142 *"the most radical innovator"* J. M. Coetzee, "The Making of William Faulkner," *The New York Review of Books*, April 7, 2005, p. 20.

142 *"the inherent tragedy of southern history"* Lorie W. Fulton, "William Faulkner's Wistaria: The Tragic Scent of the South" *Southern Studies* 11 (2004):1–9.

143 *symbol of courage and violence* William Faulkner, *The Unvanquished* (New York: Random House, 1938). True verbena should not be confused with lemon verbena (*Lippia citriodor*), which has a distinct lemony scent that Faulkner would have been hard pressed to ignore.

143 *code of honor* R. W. Witt, "On Faulkner and Verbena," *Southern Literary Journal* 27 (1994):73–84.

143 *extended use of smell* William Faulkner, *The Sound and the Fury* (New York: J. Cape and H. Smith, 1929).

146 *review was merciless* Richard Dyer, "'Blind Trust': Hold Your Nose," *Boston Globe*, June 7, 1993.

Chapter 8. Hollywood Psychophysics

147 *"audiences worldwide paid me"* Interview with John Waters, April 12, 2006.

147 *History has not been kind* Frank W. Hoffmann and William G. Bailey, *Arts & Entertainment Fads* (Harrington Park Press, 1990); Martin J. Smith and Patrick J. Kiger, *Oops: 20 Life Lessons from the Fiascoes That Shaped America*

(New York: HarperCollins, 2006); "Cinematic Stinkers," *The Times Educational Supplement,* May 26, 2006; "The 100 Worst Ideas of the Century," *Time,* June 14, 1999; Harry and Michael Medved, *The Golden Turkey Awards: Nominees and Winners, the Worst Achievements in Hollywood History* (New York: Putnam, 1980).

148   *first attempt to odorize movies*   Terry Ramsaye, *A Million and One Nights: A History of the Motion Picture* (London: Frank Cass & Co., 1926), p. 175.

148   *"tried the rose bit"*   "Kill That Butt, the Smellie Is Starting," *Film Daily,* September 10, 1958.

149   *no Rose Bowl game*   See www.TournamentofRoses.com for the relevant history.

149   *imitated by others*   Letter to the Editor by Albert E. Fowler, *Variety,* January 13, 1960 (for *Lilac Time*); *Photoplay Magazine,* September 1929, p. 98 (for *Hollywood Review of 1920*).

149   *"The scent organ was playing"*   Aldous Huxley, *Brave New World* (Garden City, NY: Doubleday, Doran & Company, 1932).

150   *"By midmorning"*   Bill Buford, *Heat: An Amateur's Adventures as Kitchen Slave, Line Cook, Pasta-maker, and Apprentice to a Dante-quoting Butcher in Tuscany* (New York: Knopf, 2006).

151   *"The blowers which wafted these odors"*   Arthur Mayer, *Merely Colossal* (New York: Simon and Schuster, 1953), pp. 187, 189–90.

151   *system described by John H. Leavell*   John H. Leavell, "Method of and apparatus for presenting theatrical impressions," U.S. Patent 1,749,187, March 4, 1930. To synchronize the smell with the movie, Leavell used a spring-loaded lever arm to detect notches cut into the edge of the motion picture film; a notch moved the lever arm and triggered the compressed air. As in Mayer's cartoon, a projectionist stood by to open valves to the appropriate odor tanks. Driven by compressed air, the scent emerged into the theater from mushroom-shaped vents beneath the seats—a standard ventilation method at the time.

151   *Walt Disney got excited*   John Canemaker, "The *Fantasia* That Never Was," *Print* 42 (1988):76–87, 139–40.

152   *"was on the verge"*   "The Smellies," *Time,* April 18, 1949, p. 30.

152   *"A young man meets"*   Felix Aeppli, *Der Schweizer Film 1929–1964: Die Schweiz als Ritual,* vol. 2, (Zürich: Limmat Verlag, 1981) p. 333 [my translation].

152 *garnering a mention* B. R. Crisler, "Week of Minor Wonders," *New York Times*, February 25, 1940.

153 *On the evening of* "Today's Program at the Fair," *New York Times*, October 19, 1940.

153 *"At the conclusion"* Hervé Dumont, *Geschichte des Schweizer Films: Spielfilme 1896–1965* (Lausanne: Schweizer Film Archiv, 1987), pp. 157–58 [my translation]. Dumont lists the first showing of *My Dream* as October 10, 1940, but also says it ran from June to July 1940. It is unclear from this whether the bust took place after the first performance or at the end of the first run.

153 *to promote his inventions* "Smellovision" item in "Sidelines" column, *Los Angeles Times*, February 3, 1946; "The Smellies," *Time*, April 18, 1949, p. 30. On February 23, 1941, the *New York Times* in its "Reported from the Field of Science" column credits Laube and Barth for inventing the "smellies." Soon after, a Swiss film encyclopedia credits them with inventing "Duftfilm"; Charles Reinert, *Kleines Filmlexikon: Kunst, Technik, Geschichte, Biographie, Schrifttum* (Einsiedeln-Zürich: Benziger & Co., 1946), p. 85.

153 *risk-taker and a feisty competitor* Michael Todd Jr. and Susan McCarthy Todd, *A valuable property: The life story of Michael Todd* (New York: Arbor House, 1983).

154 *bring aroma to movies and television* A Swiss newspaper reported on Laube's press demo of a video and smell system in a New York hotel room: "Fernriechen auf dem Weg," *Die Tat*, February 1, 1956.

155 *To secure international rights* Laube applied for a European patent in June, 1955; it was issued on January 21, 1959: Hans Laube and Bert Samuel Good, "Motion pictures and the like with synchronized odor emission," European patent GB807615. Laube filed the second U.S. application in June of 1956.

155 Motion Picture Daily *hinted* "ScentoVision to Be Installed in Theatre in 9 Months, Ruskin Says," *Motion Picture Daily*, September 14, 1956.

155 *mentioned in the* New York Times "Odors Added to Films and Video, Even Those of Oranges or Ham," *New York Times*, November 23, 1957.

156 *project called* Scent of Danger "Brand-new 'Scent' on the Todd Roster," *New York Times*, September 28, 1958.

156 *Glenda Jensen, then a secretary* Glenda Jensen, "Working for the Michael Todd Corporation and a Little Bit of Cinemiracle Too," *The 70mm Newsletter*, March 15, 2005; interview with Glenda Jensen, April 18, 2006.

156 *Elizabeth Taylor was cast*   Hedda Hopper column, *Los Angeles Times*, November 8, 1958.

156 *a public-relations executive*   "Kill That Butt, the Smellie Is Starting," *Film Daily*, September 10, 1958.

156 *revealed the cast*   " 'Does It Not Betray Itself by Its Smell?' " *Film Daily*, April 14, 1959.

157 *movie's ad slogan*   "Movies . . . Talkies . . . and Now—Smellies!" *Los Angeles Times*, April 26, 1959.

157 *odors in his machine*   Herb A. Lightman, "This Movie Has Scents!" *American Cinematographer*, February 1960, p. 92.

157 *Laube shuttled*   Interview with Carmen Laube, April 24, 2006, and with Novia Laube, April 27, 2006.

157 *on a daily basis*   Interview with Hal Williamson, April 28, 2006.

157 *needed extra time*   " 'Smell-O-Vision' to Get Film Test," *New York Times*, August 19, 1959.

157 *second U.S. patent*   "Motion pictures with synchronized odor emission," U.S. Patent 2,905,049, issued September 22, 1959; mentioned in "Times Square Conveyor System to Replace Shuttle Is Patented," *New York Times*, September 26, 1959.

158 *"rushing plans"*   "Todd 'Smell' Film May Be Scooped; Reade Rushes Own Picture to Beat 'Scent of Mystery'," *New York Times*, October 17, 1959; $300,000 is equal to $2.15 million in 2007 dollars.

159 *At a press conference*   "Todd Rival Leading in Smell-Film Race," *New York Times*, October 27, 1959.

159 *"obviously rushing to beat"*   "Scented Movies: The First Sniff?" *Newsweek*, November 9, 1959, p. 106.

159 *"the battle of the smellies"*   "Oranges Blossomed in '23 Revue," *Variety*, November 4, 1959.

159 *Installation costs* $3,500 to $7,500 equals $25,000 to $54,000 in 2007 dollars.

159 *raises a question*   Unlike Scentovision, Inc. and AromaRama Industries, Inc. which were both incorporated in New York state, there is no New York record of the Weiss Screen-Scent Corporation.

159 *"The glory that was AromaRama"*   Joan Didion, "Smellie on Seventh Avenue," *National Review*, January 30, 1960, pp. 83–84.

160   *"neither so clear nor pleasurable"*   Bosley Crowther, "On Making Scents: AromaRama Turns Out a Movie Stunt," *New York Times*, December 13, 1959.

160   *Luz Gunsberg had the same reaction*   Interview with Luz Gunsberg, June 7, 2006.

160   *"strong enough to give"*   "A Sock in the Nose," *Time*, December 21, 1959, p. 57.

160   *"quite a massive assault"*   John McCarten, "Inhalant," The *New Yorker*, December 19, 1959, p. 125.

160   *"Your clothes reeked"*   Interview with Hal Williamson, April 28, 2006.

160   *perfumer Selma Weidenfeld*   "Film Produced in Red China Turns Out to Be a Smeller," *Los Angeles Times*, December 13, 1959.

161   *"the machine-made olfactory flavors"*   "Behind the Great Wall: The sweet smell of success—via novelty, unenduring values," *Variety*, December 16, 1959.

161   *"worked part-time"*   Interview with Paul Baise, June 22, 2006.

161   *On his Christmas card*   Glenda Jensen, *The 70mm Newsletter*, 2005.

162   *only enough prints*   "Are Smellies Bottled B.O. Sunshine?" *Variety*, December 16, 1959.

163   *the entertainment included*   Judith Cass, "Recorded at Random" column, *Chicago Daily Tribune*, January 9, 1960.

163   *was received warmly*   "Diverting Tale Told with Nostril-Appeal," *Variety*, January 13, 1960; John McCarten, "Fragrant Frolic," *The New Yorker*, February 27, 1960, p. 131; "Nose Opera," *Time*, February 29, 1960, p. 98.

163   *"whole silly plot"*   Comments from Bosley Crowther in "Screen: Olfactory Debut," *New York Times*, February 19, 1960, and "How Does It Smell?" *New York Times*, February 28, 1960.

164   *"Bill got this idea"*   "An interview with Mike Todd Jr." by Roy Frumkes, posted on in 70mm.com, January 9, 2004.

165   *"a somewhat timid revolutionist"*   Hollis Alpert, *The Dreams and the Dreamers* (Macmillan, 1962), p. 179.

165   *"He was at his best"*   Michael Todd Jr. and Susan McCarthy Todd, *A valuable property*, p. 102.

165   *cynical rabbit punch*   In late November, 1959, Mike Todd Jr. announced that *Scent of Mystery* would open in Los Angeles on January 27. Reade

immediately postponed his own L.A. opening, which had been set for December 23. Later, with only a week's notice, he opened *Behind the Great Wall* at the Four Star Theater on Wilshire Boulevard on January 15—ten days earlier and two blocks down the street from Todd's film. Once again, Reade had stolen Mike junior's thunder. "Unique Film Will Screen," *Los Angeles Times*, November 24, 1959; "Musical '80 Days' Readied for Stage," *Los Angeles Times*, December 15, 1959; " 'Great Wall' Will Screen," *Los Angeles Times*, January 7, 1960.

Chapter 9. Zombies at the Mall

170   *"All around the world"*   Martin Lindstrom, *Brand Sense: Build Powerful Brands through Touch, Taste, Smell, Sight, and Sound* (New York: Free Press, 2005), p. 98.

170   *happening everywhere*   "Dollars and Scents: The Nose Knows, or Does It?" *Atlanta Journal-Constitution*, August 19, 2004; Linda Tischler, "Smells Like Brand Spirit," *Fast Company*, August 2005, p. 52; "Smells Like a Sheraton," *Washington Post*, March 5, 2006.

171   *latest in a long history*   Edward M. Ruttenber, "Sense of smell—an Important Factor in All Modern Merchandising," reprinted from the *Daily News Record* (New York) in *American Perfumer & Essential Oil Review*, June 1925, p. 208; " 'Sell by Smell' New Marketing Slogan," *Forbes*, July 1, 1934, pp. 14–15; Edward Podolsky, "Odors as sales stimulators," *The Management Review* 28 (September 1939):320; Francis Sill Wickware, "They're After Your Nose Now," *The Saturday Evening Post*, June 21, 1947, p. 26.

171   *Today's merchandisers*   "Sweet Smell of Sidewalls," *New York Times*, February 11, 2007; "A Bowling Ball with Snap (and Scent)," *New York Times*, May 6, 2007; "Scent and Sensibility," *New York Times*, September 9, 2007; "Sniff . . . and spend," *Los Angeles Times*, August 20, 2007.

172   *"perhaps the most powerful"*   "Starbucks Stirred to Refocus on Coffee," *Wall Street Journal*, February 26, 2007.

172   *"the stench of beer and sweat"*   "Luminar to Fight Smoking Ban with Sex Toys and Scent," *The Independent* (London), May 18, 2007.

172   *social psychologist Robert Baron*   Robert A. Baron, "The sweet smell of . . . helping: Effects of pleasant ambient fragrance on prosocial behavior in shopping malls," *Personality & Social Psychology Bulletin* 23 (1997):498–503.

174   *"We wanted to make"*   Bijal Trivedi, "Recruiting smell for the hard sell," *New Scientist* 2582 (December 16, 2006).

174   *chemist and physicist E. E. Free*   "Ancestral Memories in Smells," *The Literary Digest*, November 1, 1924, pp. 70–71.

174   *scientists continue to offer*   BBC News, February 19, 2004: news.bbc .co.uk/go/pr/fr/-/2/hi/technology/3502821.stm; Ann Quigley, "Smell, emotion processor in brain may be altered in depressed patients," Health Behavior News Service, press release March 10, 2003, Center for the Advancement of Health; Emma Cook, "What's Getting Up Your Nose?: These Days, If It Doesn't Smell It Doesn't Sell," *The Independent* (London); Marilyn Larkin, "Sniffing out memories of holidays past," *Lancet* 354 (1999): 2142.

175   *two equally pleasant fragrances*   A. M. Fiore, X. Yah, and E. Yoh, "Effects of a product display and environmental fragrancing on approach responses and pleasurable experiences," *Psychology & Marketing* 17 (2000):27–54.

175   *in an actual gift store*   A. S. Mattila and J. Wirtz, "Congruency of scent and music as a driver of in-store evaluations and behavior," *Journal of Retailing* 77 (2001):273–89.

175   *photos of a store*   E. R. Spangenberg, B. Grohmann, and D. E. Sprott, "It's beginning to smell (and sound) a lot like Christmas: The interactive effects of ambient scent and music in a retail setting," *Journal of Business Research* 58 (2005):1583–89.

176   *"We're Muzak for your nose"*   "Muzak Cuts Jobs; Partners with ScentAir," *Fort Mill Times*, July 7, 2005.

176   *business professor Eric Spangenberg*   E. R. Spangenberg, D. E. Sprott, et al., "Gender-congruent ambient scent influences on approach and avoidance behaviors in a retail store," *Journal of Business Research* 59 (2006): 1281–87.

176   *manipulated the scent of a mall*   J.-C. Chebat and R. Michon, "Impact of ambient odors on mall shoppers' emotions, cognition and spending: A test of competitive causal theories," *Journal of Business Research* 56 (2003):529–39.

179   *"one of those subliminal things"*   Tischler, "Smells Like Brand Spirit," p. 52.

179   *According to the psychologist Anthony Pratkanis*   Anthony Pratkanis and Elliot Aronson, *Age of Propaganda: The Everyday Use and Abuse of Persuasion* (New York: W. H. Freeman, 1992); Anthony R. Pratkanis, "The Cargo-cult Science of Subliminal Persuasion," *Skeptical Inquirer*, Spring 1992.

179   *Key—now an elderly man*   Dominic Streatfeild, *Brainwash: The Secret History of Mind Control* (New York: St. Martin's Press, 2007).

180   *German researcher Thomas Hummel*   T. Hummel, J. Mojet, and G. Kobal, "Electro-olfactograms are present when odorous stimuli have not been perceived," *Neuroscience Letters* 397 (2006):224–28.

180   *other researchers have observed*   V. Treyer, H. Koch, et al., "Male subjects who could not perceive the pheromone 5α-androst-16-en-3-one, produced similar orbitofrontal changes on PET compared with perceptible phenylethyl alcohol (rose)," *Rhinology* 44 (2006):278–82.

180   *Psychologists in the Netherlands*   R. W. Holland, M. Hendriks, and H. Aarts, "Smells like clean spirit: Nonconscious effects of scent on cognition and behavior," *Psychological Science* 16 (2005):689–93.

181   *demonstration of covert selling power*   D. A. Laird, "How the consumer estimates quality by subconscious sensory impressions; with special reference to the role of smell," *Journal of Applied Psychology* 16 (1932):241–46.

181   *a study done by some of its members*   I. E. de Araujo, E. T. Rolls, et al., "Cognitive modulation of olfactory processing," *Neuron* 46 (2005):671–79.

182   *"Unfortunately this fact offers"*   ECRO newsletter, Spring 2005, p. 6.

182   *the FCC has investigated*   FCC press statement, September 19, 2000: "The FCC's Investigation of 'Subliminal Techniques': From the Sublime to the Absurd."

182   *something experts debate*   Harper quoted in "Dollars and Scents of Business," in the *Atlanta Journal-Constitution*, June 6, 2007; Faranda quoted in "Scent: New Frontiers in Branding," in *CGI Magazine*, May 2007.

183   *"rank commercialism"*   C. Haill, " 'Buy a Bill of the Play!' " *Apollo* 126, new series 302 (1987):279–85; Calvin Trillin quoted in "Ugh, the Smell of It," *Time*, October 7, 1996.

183   *the legacy of Fred and Gale Hayman*   Steve Ginsberg, *Reeking Havoc: The Unauthorized Story of Giorgio* (New York: Warner Books, 1989), pp. 128ff, 142ff.

183   *the ScentStrip Sampler*   Everett M. Turnbull and Jack W. Charbonneau, "Fragrance-releasing pull-apart sheet," U.S. Patent 4,487,801 issued December 11, 1984.

184   *a scented full-page movie ad*   "Marketing Ploy Makes Scents," *Los Angeles Times*, September 5, 2007; Thomas Claburn, "Newspapers smell profit in scented ads," *InformationWeek*, January 29, 2007; "Joint Promotion Adds Stickers to Sweet Smell of Marketing," *New York Times*, April 2, 2007; *Angewandte Chemie International Edition*, April 27, 2007; "Scent Noses Its Ways into More Ad Efforts," *Wall Street Journal*, October 8, 2007.

184 *"Whereas you can exercise the choice"*   Emma Cook, "What's Getting Up Your Nose?" *The Independent* (London), May 16, 1999.

185 *"The television screen shows"*   A. S. Byatt, "How We Lost Our Sense of Smell," *The Guardian* online, September 1, 2001.

185 *Byatt's fiction is riddled*   A. S. Byatt, *Little Black Book of Stories* (New York: Knopf, 2004) and *The Djinn in the Nightingale's Eye* (New York: Random House, 1994).

185 *"What vile marketing decision"*   Mark Morford, "ScentStories Up Your Nose," SFGate.com, November 24, 2004.

186 *"What was once a vital instrument"*   G. G. Wayne and A. A. Clinco, "Psychoanalytic observations on olfaction, with special reference to olfactory dreams," *Psychoanalysis and the Psychoanalytic Review* 46 (1959):64–79.

186 *"Until recently, appealing to our sense of smell"*   Cook, "What's Getting Up Your Nose?"

188 *Febreze odor eliminator is equally popular*   "Sensing Opportunity in Dormitory Air," *New York Times*, January 3, 2007.

Chapter 10. Recovered Memories

189 *"Were they all collected"*   Ellen Burns Sherman, "The Redolent World," *New England Magazine* 43 (1910):319–21.

190 *"voluptuary of smell"*   Diane Ackerman, *An Alchemy of Mind* (New York: Scribner, 2004), p. 114.

190 *"great blazer of scent trails"*   Diane Ackerman, *A Natural History of the Senses* (New York: Random House, 1990), p. 17.

190 *"Proust may have been prescient"*   R. S. Herz and J. W. Schooler, "A naturalistic study of autobiographical memories evoked by olfactory and visual cues: Testing the Proustian hypothesis," *American Journal of Psychology* 115 (2002):21–32.

190 *"Proust was a neuroscientist"*   Jonah Lehrer, "The neuroscience of Proust," *Seed*, May–June 2004, p. 48–51.

190 *brand-conscious titles*   S. Chu and J. J. Downes, "Proust nose best: Odors are better cues of autobiographical memory," *Memory and Cognition* 30 (2002):511–18; S. Chu and J. J. Downes, "Long live Proust: The odour-cued autobiographical memory bump," *Cognition* 75 (2000):B41–50. For other examples, see F. R. Schab, "Odors and the remembrance of things past," *Journal of Experimental Psychology: Learning, Memory and Cognition* 16

(1990):648–55; J. A. Gottfried, A. P. Smith, et al., "Remembrance of odors past: Human olfactory cortex in cross-modal recognition memory," *Neuron* 42 (2004):687–95; A. Parker, H. Ngu, and H. J. Cassaday, "Odour and Proustian memory: Reduction of context-dependent forgetting and multiple forms of memory," *Applied Cognitive Psychology* 15 (2001):159–71; S. Chu and J. J. Downes, "Odour-evoked autobiographical memories: Psychological investigations of proustian phenomena," *Chemical Senses* 25 (2000):111–16.

191   *"This strange revival of bygone days"*   Dan McKenzie, *Aromatics and the Soul: A Study of Smells* (London: William Heinemann Ltd., 1923), p. 50.

191   *Shattuck took a close look*   Roger Shattuck, *Proust's Way: A Field Guide to* In Search of Lost Time (New York: W. W. Norton, 2000).

192   *Proust's sensory imagery*   Victor E. Graham, *The Imagery of Proust* (Oxford, England: Basil Blackwell, 1966), pp. 8, 106.

192   *poetry of Shelley and Keats*   Mary Grace Caldwell, "A Study of the Sense Epithets of Shelley and Keats," *Poet Lore* 10 (1898):573–79.

192   *"a flood of visual images"*   Graham, *The Imagery of Proust.*

192   *"I believe that odors"*   Poe, *Marginalia,* 1844.

194   *other writers were exploring*   Louise Fiske Bryson, "Training the Memory," *Harper's Bazaar,* September 1903, p. 824; "Scent and Memory," *The Spectator* (London), July 11, 1908, pp. 52–53, reprinted in *The Living Age* (Boston), November 14, 1908, pp. 437–39; "magically transported," Graham, *The Imagery of Proust,* p. 107.

194   *thoroughly psychological*   Sherman, "The Redolent World," p. 319.

194   *"These flashes of memory"*   Ellwood Hendrick, "The sense of smell," *The Atlantic Monthly,* March 1913, pp. 332–37.

195   *"The coincidence is not fortuitous"*   Charles Rosen, "Now, Voyager," *The New York Review of Books,* November 6, 1986, p. 55. According to Rosen, Proust took another writer to task for praising Ramond, specifically for praising this very passage of Ramond's.

195   *Contemporary French psychology*   Théodule Ribot, *La Psychologie des sentiments* (Paris: Félix Alcan, 1896), translated as *The Psychology of the Emotions* (London: Walter Scott Ltd., 1897), ch. 11, "The Memory of Feelings." For another example, see F. Pillon, "La Mémoire Affective: son Importance Théorique et Pratique," *Revue Philosophique* 51 (February 1901):113–38.

196   *"Sometimes, when passing through"*   Henri Piéron, "La Question de la Mémoire Affective," *Revue Philosophique* 54 (December 1902):612–15, translation by Laurence Dryer.

196   *"can only be termed ingenuous"*   Shattuck, *Proust's Way*, p. 115.

196   *sinister speculation*   Marc A. Weiner, "Zwieback and Madeleine: Creative Recall in Wagner and Proust," *MLN* 95 (1980):679–84.

197   *"The Proustian view"*   T. Engen and B. M. Ross, "Long-term memory of odors with and without verbal descriptions," *Journal of Experimental Psychology* 100 (1973):221–27.

197   *"negative experimental results"*   J. M. Annett, "Olfactory memory: A case study in cognitive psychology," *Journal of Psychology* 130 (1996):309–19.

197   *observed classic interference effects*   H. A. Walk and E. E. Johns, "Interference and facilitation in short-term memory for odors," *Perception & Psychophysics* 36 (1984):508–14; T. L. White, "Olfactory memory: The long and short of it," *Chemical Senses* 23 (1998):433–41.

198   *"the first unequivocal demonstration"*   Herz and Schooler, "A naturalistic study," pp. 21–32.

199   *"we did not find support"*   J. Willander and M. Larsson, "Smell your way back to childhood: Autobiographical odor memory," *Psychonomic Bulletin & Review* 13 (2006):240–44.

199   *criticized previous studies*   S. Chu and J. J. Downes, "Odour-evoked autobiographical memories: Psychological investigations of proustian phenomena," *Chemical Senses* 25 (2000):111–16.

199   *came a quick challenge*   J. S. Jellinek, "Proust remembered: Has Proust's account of odor-cued autobiographical memory recall really been investigated?" *Chemical Senses* 29 (2004):455–58.

200   *studies now claim*   Chu and Downes, "Proust nose best"; Willander and Larsson, "Smell your way back to childhood."

200   *A Norwegian survey*   S. Magnussen, J. Andersson, et al., "What people believe about memory," *Memory* 14 (2006):595–613.

203   *"When I was a boy"*   Haydn S. Pearson, *New England Flavor: Memories of a Country Boyhood* (New York: W. W. Norton, 1961).

203   *"A time like that"*   Ben Logan, *The Land Remembers: The Story of a Farm and Its People* (New York: Viking, 1975).

204 *"I grew up on the Nevada desert"*   Donald A. Laird, "Some normal odor effects and associations of psychoanalytic significance," *Psychoanalytic Review* 21 (1934):194–200.

Chapter 11. The Smell Museum

205 *"My collection of semi-used perfumes"*   Andy Warhol, *The Philosophy of Andy Warhol (From A to B and Back Again)* (New York: Harvest Books, 1975), p. 151.

206 *"[E]ach work of fiction"*   Bernard Benstock, "James Joyce: The olfactory factor," in *Joycean Occasions,* edited by J. E. Dunleavy, M. J. Friedman, and M. P. Gillespie (Newark: University of Delaware Press, 1991), pp. 138–56.

206 *"Cannery Row in Monterey in California"*   John Steinbeck, *Cannery Row* (New York: Viking, 1945); *Travels with Charley: In Search of America* (New York: Viking, 1962).

207 *"In the days before Prohibition"*   H. L. Mencken, *Happy Days 1880–1892* (New York: Knopf, 1940), p. 236.

207 *"there is a thick, musty smell"*   Joseph Mitchell, *McSorley's Wonderful Saloon* (New York: Pantheon, 1992; reprint of 1943 edition), p. 19.

207 *half of Americans*   "Brew a Pot? Latte Nation Thinks Not," *New York Post* online edition, August 13, 2006.

208 *manure-scented scratch-and-sniff*   "Ag Board's Brochure Is a Real Stinker," *Patriot-News,* June 17, 2005.

208 *Back in 1931*   Gove Hambidge, "Scents that make dollars; The next wave of fragrance?" *World's Work* 60 (August 1931):32–34.

208 *"I turned eight"*   Rem Koolhaas, "Singapore Songlines: Portrait of a Potemkin Metropolis . . . or Thirty Years of Tabula Rasa," in R. Koolhaas and B. Mau, *S, M, L, XL* (New York: Monacelli Press, 1995).

209 *"Now, the smell of the autumn smoke"*   Edgar Lee Masters, "Hare Drummer," *Spoon River Anthology,* 1916.

209 *"[W]e should be hanging on"*   Lewis Thomas, "On Smell," in *Late Night Thoughts on Listening to Mahler's Ninth Symphony* (New York: Viking, 1983).

210 *composition of prehistoric diets*   J. G. Moore, B. K. Krotoszynski, and H. J. O'Neill, "Fecal odorgrams: A method for partial reconstruction of ancient and modern diets," *Digestive Diseases and Science* 29 (1984):907–911.

211 *"Behind the office is a room"*   Steinbeck, *Cannery Row,* p. 22.

212   *proved too unpleasant*   Author's interview with Leti Bocanegra, August 30, 2006.

212   *Scented museum exhibits*   " 'Smellovision' Enhances Visit to Smithsonian," *Los Angeles Times,* November 30, 1967; Martin Whitfield, "Museum Haunted by a Scent of Old Times," *The Independent* (London), July 12, 1993; "Smells That Sell Not to Be Sniffed At; The T-Rex Model at London's Natural History Museum," CNN-Reuters, June 27, 2004; Matthew Tanner, "Satisfying the paying public: The effective interpretation of historic ships and boats," Third International Conference on the Technical Aspects of the Preservation of Historic Vessels, San Francisco, April 20–23, 1997.

213   *"aromatopia"*   Jim Drobnick, "Volatile Architectures," in B. Miller and M. Ward eds., *Crime and Ornament: The Arts and Popular Culture in the Shadow of Adolf Loos* (Toronto: YYZ Books, 2002).

214   *atomizer historian Tirza True Latimer*   Tirza True Latimer, *The Perfume Atomizer: An Object with Atmosphere* (West Chester, PA: Schiffer, 1991), p. 7.

214   *an exhibit honoring Gale W. Matson*   Gale W. Matson, "Microcapsules and process of making," U.S. Patent 3,516,941, issued June 23, 1970; Jack Charbonneau and Keith Relyea, "The technology behind on-page fragrance sampling," *Drug & Cosmetic Industry,* February 1997, p. 48.

215   *scent of burnt cordite*   Ad in February 1989 issue of *Armed Forces Journal International,* opposite p. 57.

216   *installation by Alex Sandover*   Henry Urbach Architecture Gallery, New York, 2000.

216   *Sissel Tolaas*   Sally McGrane, "The Odor Artist," *Wired,* April 24, 2007; also "This Art Stinks, and That's by Design," KansasCity.com/*The Kansas City Star,* February 4, 2007.

217   *"Lewis's dialectical odours"*   Drobnick, "Volatile Architectures."

218   *"It smells a little like dirty socks"*   "An Aroma Like . . . OK, OK, Plant Not Totally Foul, But No Bouquet," *Atlanta Journal-Constitution,* July 7, 1998.

219   *"Smells are surer than sounds or sights"*   Rudyard Kipling, *The Five Nations* (London: Methuen, 1903).

219   *"I saw this Australian trooper"*   A. B. "Banjo" Paterson, *Happy Dispatches* (Sydney: Angus & Robertson, 1934).

220   *three translucent globes   AIR—Urban Olfactory Installation,* in SAUMA: Design as Cultural Interface exhibit, World Financial Center, New York, June 20–September 10, 2006.

220 *accompanied a* New York Observer *reporter* Kate Kelly and Elizabeth Manus, "In a Smelly Summer, Our Team of Noses Sniffs up the City," *New York Observer,* August 9, 1999.

221 Washington Post *reporter rides along* David Segal, "Eau Dear: Sniffing Out the Big Apple's Smelliest Spots," *Washington Post,* August 17, 2006.

221 *"Yes, I admit I've taken the subway"* Paris Hilton with Merle Ginsberg, *Confessions of an Heiress: A Tongue-in-Chic Peek Behind the Pose* (New York: Simon & Schuster, 2004), p. 93.

222 *"I can easily distinguish"* Helen Keller, *Midstream: My Later Life* (Garden City, NY: Doubleday, Doran, 1929), p. 165ff.

222 *"My eyes flew open"* Celeste Bowman, "Going Home for Two," *Texas Magazine* in the *Houston Chronicle,* May 12, 1996.

222 *part of Baltimore* "You Smell That? An Olfactory-Bulb Tour of the City That Stinks," *Baltimore City Paper,* September 19, 2001.

223 *"I spent a lot of time"* www.lukeford.net/profiles/profiles/heather_ macdonald .htm

223 *"Seeping in through open windows"* "Scent of a City: Heady Essence of Oranges," *Los Angeles Times,* February 8, 2001.

Chapter 12. Our Olfactory Destiny

225 *"They were, I now saw"* H. G. Wells, *The War of the Worlds,* 1898.

227 *An informal test in 2006* "Sniffing Out Spoiled Meat," *Wall Street Journal,* December 12, 2006.

227 *a future in medicine* E. I. Mohamed, R. Linder, et al., "Predicting Type 2 diabetes using an electronic nose-based artificial neural network analysis," *Diabetes, Nutrition & Metabolism* 15 (2002):215–21; P. Dalton, A. Gelperin, and G. Preti, "Volatile metabolic monitoring of glycemic status in diabetes using electronic olfaction," *Diabetes Technology & Therapeutics* 6 (2004):534–44; "What the Nose Knows," *The Economist,* March 9, 2006.

228 *an entire book on the topic* R. Andrew Russell, *Odour Detection by Mobile Robots* (World Scientific, 1999).

228 *Amy Loutfi* M. Broxvall, S. Coradeschi, et al., "An ecological approach to odour recognition in intelligent environments," *Proceedings of the IEEE International Conference on Robotics and Automation* (ICRA), Orlando, FL, 2006;

A. Lofti, "Odour recognition using electronic noses in robotic and intelligent systems," Ph.D. thesis, Örebro University, Sweden, February 15, 2006.

228   *"in general public use"*   *Kyllo v. United States* (99–8508) 533 U.S. 27 (2001), 190 F.3d 1041, reversed and remanded.

229   *A group in Britain*   J. A. Covington, J. W. Gardner, et al., "Towards a truly biomimetic olfactory microsystem: An artificial olfactory mucosa," *IET Nanobiotechnology* 1 (2007):15–21.

230   *shred of the yeast cell membrane*   J. M. Vidic, J. Grosclaude, et al., "Quantitative assessment of olfactory receptors activity in immobilized nanosomes: A novel concept for bioelectronic nose," *Lab Chip* 6 (2006):1026–32.

230   *use bacterial cells*   J. H. Sung, H. J. Ko, and T. H. Park, "Piezoelectric biosensor using olfactory receptor protein expressed in *Escherichia coli*," *Biosensors and Bioelectronics* 21 (2006): 1981–86; Q. Liu, H. Cai, et al., "Olfactory cell-based biosensor: A first step towards a neurochip of bioelectronic nose," *Biosensors and Bioelectronics* 22 (2006):318–22.

230   *to detect explosions*   "Scent Detection Technologies Ltd. (SDT) Honoured with the 2006 Frost & Sullivan Award for Technology Innovation in the Field of Advanced Explosive Detection," PR Newswire, April 10, 2006.

230   *pushed hybridism a step further*   M. Marrakchi, J. Vidic, et al., "A new concept of olfactory biosensor based on interdigitated microelectrodes and immobilized yeasts expressing the human receptor OR 17-40," *European Biophysics Journal* 36 (2007):1015–18.

231   *better to look good*   S. A. Goff, H. J. Klee, "Plant volatile compounds: Sensory cues for health and nutritional value?" *Science* 311 (2006):815–19.

231   *production of phenylethyl alcohol*   D. Tieman, M. Taylor, et al., "Tomato aromatic amino acid decarboxylases participate in synthesis of the flavor volatiles 2-phenylethanol and 2-phenylacetaldehyde," *Proceedings of the National Academy of Sciences USA* 103 (2006):8287–92.

231   *created a tastier tomato*   R. Davidovich-Rikanati, Y. Sitrit, et al., "Enrichment of tomato flavor by diversion of the early plastidial terpenoid pathway," *Nature Biotechnology* 25 (2007): 899–901.

232   *lured out of retirement*   "Scent of a Tomato," *Sacramento Bee*, August 19, 2007.

232   *not selected for fragrance*   A. Zuker, T. Tzfira, and A. Vainstein, "Genetic engineering for cut-flower improvement," *Biotechnology Advances* 16 (1998): 33–79.

232   *biologist Eran Pichersky*   R. A. Raguso and E. Pichersky, "Floral volatiles

from *Clarkia breweri* and *C. concinna* (*Onagraceae*): Recent evolution of floral scent and moth pollination," *Plant Systematics and Evolution* 194 (1995): 55–67; E. Pichersky, J. P. Noel, and N. Dudareva, "Biosynthesis of plant volatiles: Nature's diversity and ingenuity," *Science* 311 (2006):808–11.

233   *the Fragrant Cloud rose*   I. Guterman, M. Shalit, et al., "Rose scent: genomics approach to discovering novel floral fragrance-related genes," *Plant Cell* 14 (2002):2325–38.

233   *shoot new genes*   Zuker, Tzfira, and Vainstein, "Genetic engineering for cut-flower improvement," pp. 33–79.

233   *scent of another species*   N. Dudareva, E. Pichersky, and J. Gershenzon, "Biochemistry of plant volatiles," *Plant Physiology* 135 (2004):1893–1902. See also Zuker, Tzfira, and Vainstein, "Genetic engineering."

233   *Imagine a DNA test*   A. E. Herr, A. V. Hatch, et al., "Microfluidic immunoassays as rapid saliva-based clinical diagnostics," *Proceedings of the National Academy of Sciences USA* 104 (2007):5268–73.

235   *variations in one odor receptor*   A. Keller, H. Zhuang, et al., "Genetic variation in a human odorant receptor alters odour perception," *Nature* 449 (2007):468–72.

235   *new types of consumer products*   A. N. Gilbert and S. Firestein, "Dollars and scents: Commercial opportunities in olfaction and taste," *Nature Neuroscience* 5 (2002) suppl.:1043–45.

236   *neurologist and essayist Oliver Sacks*   Oliver W. Sacks, "Dog Beneath the Skin," in *The Man Who Mistook His Wife for a Hat* (New York: Simon & Schuster/Summit, 1985).

238   *is already happening*   P. M. Smallwood, B. P. Olveczky, et al., "Genetically engineered mice with an additional class of cone photoreceptors: Implications for the evolution of color vision," *Proceedings of the National Academy of Sciences USA* 100 (2003):11706–11; Z. Syed, Y. Ishida, et al., "Pheromone reception in fruit flies expressing a moth's odorant receptor," *Proceedings of the National Academy of Sciences USA* 103 (2006):16538–43.

Index

Ackerman, Diane, 190
Acree, Terry, 27, 47
Adams, Henry, 201–4
adaptation, 84–87, 89
advertising, 170, 178, 180, 181–84, 186, 187
aging, 54–55
air fresheners, 106, 144–45, 185–86
AIR—Urban Olfactory Installation, 220
alcohol, 72
Alexandrov, Grigory, 152
Allison, John, 218
Allure magazine, 184
Alpert, Hollis, Jr., 163, 165
Amoore, John, 21–22
Amorphophallus titanum, 218–19
amygdala, 174
androstadienone, 235
androstenone, 235
Angewandte Chemie, 184
animals, 64, 65
Annett, Judith, 197
anosmia, 49–51, 55, 234
apes, 64, 65
Appelo, Tom, 130
Aroma Jockey, 131
aroma models, 45–46
AromaRama, 147, 148, 159–68
aromas, 96, 99, 102, 104, 112, 135, 173, 210
aromatherapy, 90, 188

aroma wheels, 9–10
Around the World in 80 Days (film), 155
art, 215–17
Arthur D. Little, Inc., 3
artifical olfactory mucosa, 229–30
atomizer, 213–14
audio tapes, 179–80
Autocrat of the Breakfast Table, The (Holmes), 193–94
aversions, 114, 116–17, 119
awareness, 128–30
  conscious, 178
Axe body spray, 188
Axel, Richard, 2, 65

babies, 29–30
bacon, 96–97
bacteria, 99
Baise, Paul, 161, 162
Baron, Robert, 172–74
Barry, Dave, 53
Barth, Robert, 152, 153
Beard, James, 96
Beautiful (perfume), 15
beer, 8–9, 172
Beer Flavor Wheel, 8–9
behavior, 172–74, 181
  helping, 172–73
  of consumers, 173–5, 182
Behind the Great Wall (film), 159–62, 270n.165

Belasco, David, 132
Bell, Rolf, 114
Bellenson, Joel Lloyd, 107–8
Belock Instrument Company, 157–58
Benjamin, Robert, 157
Benstock, Bernard, 206
Berghammer, Eric, 131
Bergson, Henri, 196
Bhatnagar, Kunwar, 64
Billing, Jennifer, 99
bladder cancer, 62
Blagojevich, Rod, 30
blast injection, 77–78
blindness, 56, 57
Blind Trust, 144–46, 148, 150
body odor, 52, 182
Book of Perfumes, The (Rimmel), 119
Boring, Edwin, 3
Boston Globe, 146
Bowman, Celeste, 222
brain, 24, 89, 114, 194
    adaptation process, 84
    and emotional responses, 174
    as interpreter, 26, 64, 66
    management of odor, 80
    and molecules, 25
    odor particles in, 75
    and odor perception, 53, 55, 76, 90
    and olfactory complexities, 23
    pattern of response, 67–68
    and smell loss, 55
    subconscious registration of odors in,
        180–81
    tumors, 76, 77
Brand Sense (Lindstrom), 170
Brave New World (Huxley), 149
Breathe Right nasal dilator, 83, 84
Breslin, Paul, 54
Brewer's Clarkia, 232
Bright Lights, Big City (McInerney), 135
Brill, A. A., 58, 59
Brown, Donald E., 105
Brush, Sue, 178

Bryson, Louise Fiske, 194
Buck, Linda, 2, 65
Bucknell University, 21
Buford, Bill, 150
Burr, Chandler, 17
butterflies, 39–40
    Green-veined White, 40, 41
    Mustard White, 40
Byatt, A. S., 184–85

cacosmia, 52
caffeine, 109
Caldwell, Mary Grace, 192
California, 223–24
Calkin, Robert, 10, 67
cannabis, 72
Cannery Row (Steinbeck), 206, 211
Carbonnières, Louis-François
        Ramond de, 195
Cargille Scientific, Inc., 21
Carles, Jean, 11
Cartman, Eric, 103
casinos, 171
Castle, William, 166
categorical perception, 5
Cather, Willa, 141–42
cerebellum, 80, 81
Chamaebatia. See Sierra Mountain
        Misery
Chebet, Jean-Charles, 176
cheese, 107, 109
Chekhov, Anton, 134
chimpanzees, 66, 95, 100, 101
Chirac, Jacques, 103
Chu, Simon, 199–200
Cinerama, 154, 155, 161
Cinestage Theatre (Chicago), 167
citric acid, 94
Clark, Larry, 66
clichés, 134
climate, 99, 100
Clinco, A. A., 186
Cobain, Kurt, 130

cocaine, 31–32, 63
coffee, 46, 207
coffee beans, 107–9, 171–72
cologne, 10, 12–13, 67, 205–6
color, 4–5, 94, 127
congruency, 174–75, 176, 177
conscious awareness, 178
context, 175, 176
Cook, Emma, 184, 186
cooking, 95–100, 207
coprolite, 210
corpses, 121–25
creativity, 132, 136–37
cribriform plate, 75
Crocker, Ernest C., 3, 4, 20, 108
Crocker-Henderson system, 21
Cross, Charles, 130
Crouch, Melissa, 131
Crowther, Bosley, 163
cultures, 97–99, 103–4, 234

Daily News Record, 171
Dalton, Pam, 54, 88–89, 119
Darwin, Charles, 63, 128–29
Davis (Calif.), 5–6
DeMille Theater (N.Y.), 159
Dick, Philip K., 52
Dickinson, Emily, 137–39
Didion, Joan, 159
DigiScents, Inc., 148
Disney, Walt, 151–52
Divakaruni, Chitra Banerjee, 70
DNA, 100, 230, 233, 234
dogs, 61–62, 65, 66, 100, 102
    and bladder cancer, 62
    and drugs, 31, 32, 63
Doll, Bill, 156, 158, 164
Dong, Loc, 217
Dostoyevsky, Feodor, 134
Doty, Richard, 72, 73
Downes, John, 199–200
Dravnieks, Andrew, 222
Drobnick, Jim, 213, 215, 217

Dröscher, Vitus, 2
drug smells, 70–72
drunk driving, 72
Duchafour, Bertrand, 220

Eames, Charles, 133
Eames, Ray, 133
eating, 50–51
electronic nose, 226–31
Ellis, Havelock, 64, 92
Elsberg, Charles A., 76–78, 81
emotions, 174, 176, 178, 198–200
empathy, 128, 130–31
Engen, Trygg, 2, 3, 127, 197
essential oils, 42
eucalyptus, 42
European Chemoreception Research
    Organization, 181–82
experts, 67
extinction, 117
extinct smells, 206–11

Faranda, Joe, 182
Faulkner, William, 142–44
Feather River, 35, 36, 39
fecal odor, 28, 30, 112, 210
Federal Communications
    Commission, 182
Feynman, Richard, 63
films, 147–69, 266n.151
Finck, Henry T., 92–93
flatus, 28–29
flavor, 91–92
flavor principles, 97–98, 100
Fletcher, Harvey, 214
Fliess, Wilhelm, 58, 59, 60
floral scent biochemistry, 232
flowers, 137–39, 232–33
Flynt, Larry, 215
fMRI brain imaging, 134
food, 50–51, 92, 94–99, 102, 104, 210
Forbes, 171
Fox, Kate, 174

Fox-Walden Films, 184
France, 106–7
Frank, Bob, 82
Freud, Sigmund, 58–60
fruit juices, 93
fungi, 99

Galbraith, John Kenneth, 129
garlic, 99, 103
gas chromatography, 26–28, 36,
    37, 38, 210
gas chromatography-olfactory
    (GC-O), 27, 43
Gay, Peter, 58
gender, 29, 52–54, 176
genetics, 231–35
gene-transfer technology, 237
genomics, 101, 234–35
genotype, 234
Gesteland, Bob, 82
Gibson, William, 237
Giorgio (perfume), 183–84
Givaudan method, 11
Givaudan-Roure Fragrances, 144, 146
Goethe, Johann Wolfgang von, 137
Good, Bert, 157
Good Earth, The (Buck), 103
Graham, Victor, 192
grapefruit, 175
Great Britain, 212
Gulbrandsen, Marc, 167
Gunsberg, Luz, 160
Gunsberg, Sheldon, 160

Haarmann & Reimer, 13
Hamlet (Shakespeare), 111
Harper, Michelle, 182
Hawthorne, Nathaniel, 135–36,
    192
Hayman, Fred, 183
Hayman, Gale, 183
head injury, 50
headspace capture, 37

hearing, 5, 64
heirloom tomatoes, 232
Henderson, Lloyd F., 3, 20, 108
Hendrick, Ellwood, 194
Henning, Hans, 19–20
Hepper, Peter, 61
Herz, Rachel, 174, 190, 198
hexenone, 38
Heymann, Hildegaarde, 6, 7, 109
Hinton, Devon, 116
Hirst, Damien, 75
Holmes, Oliver Wendell, 193–94
homesickness, 219
hominids, 95
hormones, 53–54
House of Seven Gables, The
    (Hawthorne), 135–36, 192
Howes, David, 59
Howland, William, 133
human genome, 101, 234
humans, 100–101
Hummel, Thomas, 180
Huntley, Chet, 161
Huxley, Aldous, 149
hybridism, 230
hyposmia, 49, 60

Idiopathic Environmental Intolerance,
    113–14, 118–19
illness, 112, 227
imagination, 128, 131–32
infectious disease, 49
interference, 205
interference effects, 198
Internet, 148
introspection, 195

Jellinek, Stephan, 10, 67, 199–200
Jensen, Glenda, 156
Jha, Radhika, 104
Jitterbug Perfume (Robbins), 70
Johns, Elizabeth, 197
Jones, Harris, 33

Kaiser, Roman, 36–37, 38, 43
Keats, John, 192
Keillor, Garrison, 105, 202
Keller, Helen, 56–57, 70, 73, 222, 223, 224
Kellwaye, Simon, 112
Kemp, Sarah, 131
Key, Wilson B., 179
kimchi, 105
Kipling, Rudyard, 219
KMEZ-FM, 182
Knasko, Susan, 88
Koolhaas, Rem, 208
Kozári, Hilda, 220
Kuklinski, Richard, 122

Ladder of Years (Tyler), 134
Lagerfeld, Karl, 216
Laing, David, 23–24, 78–80
Laird, Donald, 181
Lake Wobegon Days (Keillor), 202
Larsson, Maria, 199
laryngectomy, 82–83
Laska, Mathias, 64–65
Las Vegas (Nev.), 171
Latimer, Tirza True, 214
Laube, Carmen, 167–68, 169
Laube, Hans E., 152–57, 159, 164, 167–69
Laube, Novia, 168–69
lavender, 175
Lawless, Harry, 127
leaf burning, 209
learned aversion, 116–17
Leavell, John H., 151, 266n.151
Le Guérer, Annick, 59
Lehrer, Jonah, 190
lemon, 98, 99
Lewis, Mark, 217
licorice, 210–11
Lily of the Valley (fragrance), 175
linalool, 232
Lindstrom, Martin, 170–71, 188

Linnaeus (Carl von Linné), 18–19
literature, 133–44, 185, 189–94, 206
    See also specific literary works
Logan, Ben, 203
Longstaff, George, 40
Loufti, Amy, 228
Love for Three Oranges, 133
Luminar (co.), 172
lutefisk, 104–5

MacDonald, Heather, 223
magazine ads, 183–84
Magnificent Ambersons, The (Tarkington), 209
Major Histocompatability Complex, 61
Mali, Sandrine, 216–17
malls, 172–73
malodor, 111–25
Management Review, The, 171
manure, 208
marijuana, 30, 32–34, 70–73, 228
Marin County (Calif.), 111, 112
marketing, 170–88
Martians, 225
masochism, 140
mass spectrometer, 27, 36, 37, 38
Masters, Edgar Lee, 209
Matson, Gale W., 214
Matter and Memory (Bergson), 196
Maybach, Wilhelm, 214
Mayer, Arthur, 150–51
McAfee, Helen, 134
McCarten, John, 164
McCormick & Company, 184, 222
McInerney, Jay, 135
McKenzie, Dan, 191
McNamara, Mary, 223
MCS. *See* Multiple Chemical Sensitivity
McSorley's Tavern (N.Y.), 207
meat, 95–96, 99
Meilgaard, Morten, 8–9

memory, 55, 189–204
Mencken, H.L., 206–7
metaphor, 142
methyl benzoate, 31–32, 63
methyl salicylate, 41
mice, 65–66, 100, 150
Michon, Richard, 176
Midnight's Children (Rushdie), 69
Millikan, Robert A., 214
Mistress of Spices, The (Divakaruni), 70
MMP
    4-methylpentan-2-one, 43
molecules, 25–28, 38, 43–44, 65, 75,
    76, 226, 234–36
Monardella, 35
monkeys, 65
mood, 51, 90, 173–74, 176
Morford, Mark, 185–86
Morgenthaler, René, 11
movies. *See* films; *specific movies*
Muir, John, 35–36, 39
multimedia, 133
Multiple Chemical Sensitivity (MCS),
    112–13, 118–19
multisensory branding, 170
Murphy, Michael, 1–2
museum exhibits, 211–12
music, 175
musk, 128, 234
mutations, 101, 235
Muzak LLC, 175
My Dream (film), 152–53, 154

Nabokov, Vladimir, 39
Nakamura, Lynn, 123
nasal airflow, 76, 83
nasal vestibule, 84
National Geographic Smell Survey,
    49, 55
National Institutes of Health, 50
National Steinbeck Center, 211
natural botanical scents, 41–42, 47
nerve cells, 64, 66, 75

Neuromancer (Gibson), 237
New Line Cinema, 166
New York City, 220–22
New York death, 120–21
New Yorker, The, 160, 161
New York Times, 17, 77, 132, 152–53,
    157, 160, 161
Noble, Ann, 6, 7, 232
nose
    access to amygdala, 174
    electronic, 226–31
    filters, 235–36
    receptor genes, 101
    sensory functions of, 26, 64, 237
    as site of smell, 75
    and smell loss, 55

Odorama, 147, 166
Odorated Talking Pictures, 152, 154
odor perception
    in animals, 65
    and brain, 53, 55, 76, 90
    cause of, 25–26
    degradation of, 101
    enhancement of, 236
    experiments, 61, 88
    Freud on, 58
    genetics of, 234–36
    and mutations, 235
    neuroanatomical basis of, 193
    pathologies of, 51–52
    psychology of, 93
    research into, 50
    and smoking, 55–56
    and sniffing, 81–82, 83
    and spin, 89–90
    subliminal, 180
    variability of, 48, 233–34
    Zwaardemaker's device for, 77
odor prism, 19–20
odors
    adaptation to, 84–87
    blanding, 106–7

boosters, 236
capture of, 36–37
classifications of, 18–23
as emotional, 174, 176
extinct, 206–11
and flavor, 92
Freud on, 58–60
hallucinations of, 51
identification tests, 49
and memory, 189–204
number of, 1–4
phobias to, 116–19
popular, 208
primary, 21–22
receptors, 65–66, 101, 237
and sight, 93–94
subliminal, 178–83
and taste, 94
tests of, 49, 82, 85
vocabulary for, 126–27
See also Odor perception; Sniffing;
   specific odors
olfactometer, 77, 78
olfactory cleft, 75, 76
olfactory experts, 67
olfactory genius, 128–36
olfactory neurochip, 230
olfactory reference syndrome, 52
olfactory sensitivity, 48, 54, 76–77,
   78, 113
olfactory suggestion, 87–88
O'Mahony, Michael, 88
onions, 99
O Pioneers! (Cather), 141–42
orange tree, 42
orbitofrontal cortex, 67
O'Rourke, P.J., 105
Osmothèque (Versailles, France), 213
ovulation, 59

Paglia, Camille, 138, 139
Panavision, 154
panic attacks, 116

Parkinson's disease, 83
parosmia, 52
Parylene C., 230
patchouli oil, 34, 235
Pause, Bettina, 174
Payne, Jerry, 122–23
Pearson, Haydn, 203
peppers, 99
perceptual skills, 67
Perfume: The Story of a Murderer
   (Süskind), 68–69, 130
perfumery, 10
   accords, 12, 13, 97
   activists against, 112–13
   coffee beans as reset button,
      107–8
   computerization, 12
   creation of perfume, 12
   critics, 16–17
   fragrance families, 14
   Givaudan method, 11
   guides, 13–14
   Ingredient Voice/Imagery Voice,
      15–16
   museum, 213
   olfactory fatigue in, 85
   professional researchers, 67
phantosmia, 51
phenotype, 234
phobias, 116–19
Pichersky, Eran, 232–33
Piéron, Henri, 196
place, 219–20
placebos, 88, 89–90
plants, 18, 233
Poe, Edgar Allan, 192
Polyester (film), 147, 166, 215
polymers, 226, 230
Pratkanis, Anthony, 179, 180
prehistoric diets, 210
primary odors, 21–22
primates, 65, 100–101
print advertising, 183–84

Proust, Marcel, 189–92, 194–202, 204
psychogenic theory, 118–19
pubs, 172, 207

quartz crystal microbalance, 230

Ralston, Aron, 125
Ramsland, Katherine, 122
"Rappaccini's Daughter" (Hawthorne),
    136
rats, 65, 100
Reade, Walter, Jr., 158–59, 161, 162,
    164–68, 269–70n.165
receptor genes, 65–66, 100–101,
    234, 237
rehearsal effects, 198
Remembrance of Things Past (Proust),
    190–92
retronasal olfaction, 92–93, 102
Revue Philosophique, 195–96
Rhodia (co.), 156, 160
Rialto Theater (N.Y.), 151
Ribot, Théodule, 195
Rice, Jerry, 84
Rimmel, Eugene, 119, 183, 186
Robbins, Tom, 70
Roos, Audrey, 156, 158
Roos, William, 156, 158
rose maroc, 176
Rosen, Charles, 195
roses, 232–33
Rothafel, Samuel "Roxy," 148
Rozin, Elisabeth, 97, 98, 100
Rozin, Paul, 59
Rugrats Go Wild (film), 166
Rushdie, Salman, 69

Sabini, John, 88
Sacher-Masoch, Leopold von, 140–41
Sacks, Oliver, 236
Safari (perfume), 85–86
Sagan, Carl, 92
St. James, Izabella, 114
Sales-Girons, Jean, 213

Sandover, Alex, 216
Saturday Evening Post, The, 171
Scarlet Letter, The (Hawthorne), 136
Scent-Air Technologies, Inc., 175
scented entertainment, 144–69
scented print advertising, 183–84
scent marketing, 170–88
scent masking, 114–15
Scent of Mystery (film), 156, 162–65
Scentovision, Inc., 155
scent tracking, 102
Schiaparelli (co.), 158
Schiller, Friedrich, 136–37
Schlaepfer, Conrad A., 152, 153
Science Digest, 92
Scotland, 172
scratch-and-sniff, 214–15, 216
Sea Mist (fragrance), 175
search and seizure, 70–71
seasonings, 98
semantic profiling, 22
Sephora stores, 14
Serrano, Andres, 216
sewers, 112
sex, 51, 59, 140–41, 236
Shattuck, Roger, 191, 196
Shelley, Percy Bysshe, 192
Shepherd, Gordon, 64, 102
Sherman, Ellen Burns, 189, 194
Sherman, Paul, 98–99, 100
shopping malls, 172–73
short-term memory, 55
sick building syndrome, 90
Sierra Mountain Misery, 35, 36,
    37–38, 39
Simpson, O. J., 121
Simulacra, The (Dick), 52
Singapore, 208
single nucleotide polymorphisms, 235
Slosson, Edwin E., 87, 88
Smell (Radhika), 104
smell loss. *See* anosmia
Smell-O-Vision, 147, 148, 152,
    156–59, 162–69

smells. *See* odors
Smith, Dexster, 107–8
Smith, Timothy, 64
smoking, 55–56, 172
sniffing, 74–85
    cerebellum's role in, 80
    of imaginary smell, 81
    and odor perception, 81–82, 83
    physical characteristics of, 78
    purpose of, 75
    remedial, 82–87
    shampoos, 177
    types of, 79
Sobel, Noam, 80, 81
Sound and the Fury, The (Faulkner),
    143–44
Soviet Union, 152
soy sauce, 94
Spangenberg, Eric, 176
Speak, Memory (Nabokov), 39
Spectator, The, 194
spices, 97–100, 184
spin, 89–90, 118
Stanley Warner Corporation, 154
Starbucks Coffee, 171–72
Steinbeck, John, 206, 211–12
stereotyping, 103
Stevenson, R.J., 94
stimulus generalization, 117
strawberry, 94
subliminal scents, 178–83
Subliminal Seduction (Key), 179
sucrose, 94
suggestion, 87–88, 89, 182
surprise, 200–201
Surströmming, 104
Süskind, Patrick, 68, 130
Swann's Way (Proust), 190–91
symptom learning, 117
systematic desensitization therapy, 117

Tarkington, Booth, 209
taste, 91–92, 94
taverns, 206–7

taxonomy, 18
Taylor, Elizabeth, 155, 156, 157, 158,
    163, 165
Tec, Roland, 144–46
television, 153
terpenes, 32
theatrical scent, 132–33
This Is Cinerama (film), 154, 161
Thomas, Lewis, 209
Thomas, Lowell, 161
threshold detection tests, 49
Tiger, Lionel, 54
Time magazine, 160, 161, 164
Tingler, The (film), 166
tip-of-the-nose phenomenon, 127
Todd, Michael, 153–55, 161, 165, 168
Todd, Mike, Jr., 155–57, 159, 161,
    163–66, 169
Todd Organization, 158
Tolaas, Sissel, 216
tomatoes, 231–32
Toot Trapper, 29
tracheostomy valve, 83
trauma, 115
Trillin, Calvin, 183
Trotter, Charlie, 98
Turin, Luca, 16
Tyler, Anne, 134

Ulysses (Joyce), 206
Une Odeur de luxe, 217
United Artists, 156, 157
University of California, Davis, 6
Unvanquished, The (Faulkner), 142–43

Van den Bergh, Omer, 116–17, 118,
    119
vanilla, 176
Vanity Fair, 183
van Toller, Steve, 174
Variety, 159, 161, 162, 163
vegetables, 100
verbal overshadowing effect, 67
verbal skills, 53

Vicary, James, 179
visual imagery, 81, 131–32
Vitarama, 154
volatiles, 45, 46, 47, 227
Vongerichten, Jean-Georges, 217

Wade, Nicholas, 101
Wagner, Richard, 139–40, 196–97
Wakabayashi, Dennis, 123, 124
Wales, 172
Walk, Heidi, 197
Walker, Herschel, 83
Waller, Fred, 153–54
Warhol, Andy, 30, 205
Waters, John, 147, 166, 215
Wayne, G. G., 186
Weidenfeld, Selma, 160–61
Weiner, Marc, 139, 196–97
Weiss, Charles, 156, 159, 162
Weiss Screen-Scent Corp., 156
Wells, Deborah, 61
Wells, H. G., 225
Wesendonk, Mathilde, 196

White, Theresa, 198
White Tea (fragrance), 178–79
Whitman, Walt, 38–39, 74, 238
Willander, Johan, 199
Williamson, Hal, 157, 160, 162, 165
wine, 6–8, 67, 68, 109–10
Wine Aroma Wheel, 7–8
wintergreen, 41
Wolfe, Tom, 206
women, 52–54, 59
wooden sticks, 62–63
Woodford, W. James, 31–32, 33,
    71–72
words, 134
Wrangham, Richard, 95, 96
Wright, R. H., 2–3

Yamazaki, Kunio, 61
young people, 187–88

Zellner, Debra, 93
Zola, Emile, 129
Zwaardemaker, Hendrik, 19, 77